Data Eng[illegible]g with Medallion Architecture

Building scalable multi-cloud pipelines with auditable governance and automated DevOps

Miki Eto

bpb

www.bpbonline.com

First Edition 2026

ISBN: 978-93-65894-233

LIMITS OF LIABILITY AND DISCLAIMER OF WARRANTY

Dedicated to

My parents, Tsutomu and Kimie Eto,
my siblings, their families and all who came before me,
whose love and lifelong efforts made this book possible

About the Author

Miki Eto has a bachelor's degree in mathematics. He has over 15 years of experience in enterprise data solutions, including data platform engineering, cloud architecture, data integration, ETL/ELT pipelines, and technical pre-sales. He has led data platform modernization projects at Fortune 500 companies in the insurance and consumer goods industries, implementing data mesh and medallion architectures. He established governance frameworks with unified change flows, rollback capabilities, and audit trails, the same methodology presented in this book. He is fluent in Japanese and English.

About the Reviewers

- **Dr. Michele Furlan** is a technology leader working at the intersection of data engineering, data science, and applied AI. Over the years, he has led multidisciplinary teams to design and build large-scale data platforms and production-grade AI systems in complex, highly regulated industries. He holds a graduate certificate in mathematics, an MSc. in psychology, and a Ph.D. in cognitive neuroscience.

  He is currently working as head of AI, and he leads the development, deployment, and governance of AI solutions.

- **Ethan** is a lead data engineer specializing in Databricks and is a certification subject matter expert. He has deep knowledge of medallion architecture and Spark. He is passionate about all things data and AI and how they can contribute to greater societal good. He is based in Belfast, Ireland.

Acknowledgement

I would like to thank the technical reviewers whose detailed feedback strengthened every chapter of this book. Their insights on architecture, governance, and real-world implementation challenges were invaluable.

My sincere gratitude goes to the team at BPB Publications. Their guidance, patience, and thoughtful feedback throughout the writing process made this book possible.

I would also like to thank my colleagues and friends, past and present, who have shaped my understanding of data engineering through countless discussions, code reviews, and late-night debugging sessions.

Finally, I thank my family for their continued encouragement.

Preface

It is 2 AM. Production is down. A schema change deployed yesterday broke everything.

Your team scrambles. Someone suggests rolling back, but no one knows if it is safe. The audit log is scattered across three systems. The engineer who approved the change is on vacation. Four hours later, you are back online, but the CEO wants answers.

This book exists so that it never happens again. Modern data platforms face a fundamental challenge: how do you organize data so that raw ingestion, transformation, and business-ready outputs remain cleanly separated yet seamlessly connected? The medallion architecture, with its bronze (raw), silver (cleansed), and gold (business-ready) layers, provides a proven pattern adopted by organizations worldwide.

However, architecture alone is not enough. Without governance, even the best-designed medallion layers become fragile. A schema change in bronze can cascade to gold. A failed transformation can corrupt months of curated data. This book teaches you not just how to build medallion architecture, but how to operate it safely in production.

Throughout this book, you will follow Fusion-Mart, a fictional online retailer tackling inventory discrepancies, real-time inventory updates, fraud detection, and cloud migration. Chapters 2 through 10 include hands-on labs where you practice these concepts and produce evidence files that mirror production implementations.

This book is divided into 10 chapters. They cover the foundational architecture, measurable business value, scalable pipelines, comprehensive governance, AI/ML deployment, and platform optimization needed to build and operate production-grade data platforms. The details are listed in the following:

Chapter 1: Evolution of Data Architecture- This chapter covers why governance matters and reveals lessons from architectural failures. You will understand how ungoverned data platforms lead to cascading failures and technical debt.

Chapter 2: Understanding Data Mesh, Lakehouse, and Medallion- This chapter provides the framework for defining boundaries and selecting the right architecture pattern for your organization.

Chapter 3: Data Integration Strategy, Business Impact, and ROI- This chapter equips you with SLIs, SLOs, and evaluation methods to prove your platform's business impact.

Chapter 4: Medallion Architecture in Multi-cloud- This chapter implements bronze, silver, and gold layers across cloud providers with consistent security and governance.

Chapter 5: Building Scalable Data Pipelines- This chapter gives you Kafka, dbt, and Terraform pipelines with guaranteed rollback capabilities.

Chapter 6: Data Governance and Compliance- This chapter establishes data vault modeling and consolidated audit trails for regulatory requirements.

Chapter 7: MLOps for AI Model Deployment and Monitoring- This chapter demonstrates how to deploy and monitor AI models with the same rigor as human-authored code.

Chapter 8: DevOps and CI/CD for Data Engineering- This chapter explores the automated governance checkpoints through CI/CD pipelines.

Chapter 9: Cloud Migration and Coexistence Strategies- This chapter covers coexistence patterns for live migrations without disrupting operations.

Chapter 10: Scaling Data Platforms with Optimization- This chapter delivers performance tuning and cost optimization techniques.

This book represents a fundamental change in how we think about data platform governance. The old model asked *Who wrote this code?* and trusted senior engineers while scrutinizing junior ones. The new model asks, *Is this change safe?* regardless of the author. Gates do not care whether a change came from a senior architect with twenty years of experience or an AI assistant that generated code in seconds. They evaluate the same criteria: Does it respect boundaries? Can we roll back? Is there evidence? Every change earns its way to production through the same objective evaluation.

Welcome to the future of data engineering.

Code Bundle and Coloured Images

Please follow the link to download the
Code Bundle and the ***Coloured Images*** of the book:

https://rebrand.ly/wa6do5j

The code bundle for the book is also hosted on GitHub at
https://github.com/bpbpublications/Data-Engineering-with-Medallion-Architecture.
In case there's an update to the code, it will be updated on the existing GitHub repository.

We have code bundles from our rich catalogue of books and videos available at https://github.com/bpbpublications. Check them out!

Errata

We take immense pride in our work at BPB Publications and follow best practices to ensure the accuracy of our content to provide an indulging reading experience to our subscribers. Our readers are our mirrors, and we use their inputs to reflect and improve upon human errors, if any, that may have occurred during the publishing processes involved. To let us maintain the quality and help us reach out to any readers who might be having difficulties due to any unforeseen errors, please write to us at: errata@bpbonline.com

Your support, suggestions and feedback are highly appreciated by the BPB Publications' Family.

Piracy

If you come across any illegal copies of our works in any form on the internet, we would be grateful if you would provide us with the location address or website name. Please contact us at business@bpbonline.com with a link to the material.

If you are interested in becoming an author

If there is a topic that you have expertise in, and you are interested in either writing or contributing to a book, please visit www.bpbonline.com. We have worked with thousands of developers and tech professionals, just like you, to help them share their insights with the global tech community. You can make a general application, apply for a specific hot topic that we are recruiting an author for, or submit your own idea.

Reviews

Please leave a review. Once you have read and used this book, why not leave a review on the site that you purchased it from? Potential readers can then see and use your unbiased opinion to make purchase decisions. We at BPB can understand what you think about our products, and our authors can see your feedback on their book. Thank you!

For more information about BPB, please visit www.bpbonline.com.

Join our Discord space

Join our Discord workspace for latest updates, offers, tech happenings around the world, new releases, and sessions with the authors:

https://discord.bpbonline.com

Table of Contents

Reading Guide

Introduction

This guide helps you navigate the book effectively. Whether you read cover to cover or jump to specific chapters, use this section to understand how the pieces fit together.

The *Preface* introduces the three questions every data team must answer and explains why governance matters. This *Reading Guide* shows you how to apply that knowledge: it organizes the ten chapters into six key areas (architecture, business value, pipelines, governance, AI/ML, and scale), maps them to the Four-Gate Governance Model, and explains the hands-on labs where you practice each concept. Think of the Preface as your motivation and this guide as your roadmap.

Use this guide to identify which chapters address your current challenges, understand how the Four-Gates apply across all areas, and plan your learning path based on your team's priorities.

Navigating this book

This book is organized into six key areas, as listed in the following table:

Area	Chapters	Focus	Primary gate
Architecture	1, 2	Evaluate data mesh, lakehouse, and medallion patterns	Boundary gate
Business value	3	Connect data quality to ROI	
Pipelines	4, 5	Build systems that survive production failures	RB-30 gate
Governance	6, 8	Answer auditors in one click	Evidence gate
AI/ML	7	Deploy AI models with the same rigor as human code	AI proposal gate
Scale	9, 10	Migrate and optimize without breaking production	

***Table 1**: Six key areas mapped to primary governance gates*

If you are new to medallion architecture, start with *Chapter 1, Evolution of Data Architecture,* and *Chapter 2, Understanding Data Mesh, Lakehouse, and Medallion*. If you are focused on a specific challenge, such as audit compliance or AI governance, use the table above to find the relevant chapters.

Three questions you must answer

Before discussing implementation, every data platform team must answer these three strategic questions:

- **Question 1: Can we trust AI-generated proposals?**

  As AI systems increasingly generate infrastructure code, data pipelines, and schema changes, organizations need objective criteria to evaluate these proposals before they reach production. *Chapter 7, MLOps for AI Model Deployment and Monitoring,* addresses this through the AI proposal gate.

- **Question 2: How do we guarantee production safety?**

  When a deployment fails, your team needs confidence that they can restore service quickly. Without proven rollback capabilities and clear boundaries, every change becomes a gamble. *Chapter 2, Understanding Data Mesh, Lakehouse, and Medallion,* and *Chapter 5, Building Scalable Data Pipelines,* establish the boundary gate and RB-30 gate to answer this.

- **Question 3: How do we answer auditors?**

  When compliance asks *who approved this change and why,* you need evidence in minutes, not days. Without consolidated evidence, compliance audits become multi-day excavations through scattered logs. *Chapter 6, Data Governance and Compliance,* builds the evidence gate for this purpose.

If you cannot confidently answer *yes* to all three, your platform is at risk. The Four-Gate Governance Model addresses each question systematically.

The Four-Gate Governance Model

The Four-Gates translate these strategic questions into tactical checkpoints. Every change, human or AI-generated, must pass through all four gates before reaching production and is listed in the following table:

Gate	Validation	Answers	Chapter
Boundary	Change respects constraints (budget, security, schema)	Question 2	2
RB-30	Change can be rolled back within 30 minutes	Question 2	5
Evidence	Change has a complete audit trail (who, what, why)	Question 3	6
AI proposal	AI-generated change meets the same standards as human code	Question 1	7

***Table 2:** Four governance gates validation criteria and chapter references*

Each gate produces one of three outcomes: **Go** (proceed), **Retry** (fix and resubmit), or **Kill** (reject). This framework eliminates 3 AM escalations, fails a gate, fixes the issue, resubmits, and ships.

Hands-on labs

Chapters 2 through 10 include hands-on labs in GitHub Codespaces. Every lab produces a **`result.json`** file with this structure as listed in the following table:

Chapter	Lab name
2	Boundary and RB-30 gate validation
3	Integration SLO/SLI check
4	Medallion consistency check
5	Mini single change highway
6	Data governance with tiny data vault
7	AI-generated change pack evaluator
8	DevOps and collaboration on a tiny CI/CD pipeline
9	Cloud migration and coexistence
10	Scaling and optimizing

Table 3: *Hands-on labs and their learning objectives by chapter*

`result.json` format, as listed in the following table:

Field	Purpose
status	Accept or reject the decision
checks	Validation tests passed or failed
metrics	Quantitative measurements
messages	Human-readable explanations

Table 4: *Standardized result.json format for evidence collection*

Note: **Labs use accept/reject for simplicity. In the book text, we use Go/Retry/Kill to describe the same decision workflow.**

Fusion-Mart's journey

Throughout the book, you follow Fusion-Mart, a fictional online retailer demonstrating how the Four-Gate Governance Model solves real-world challenges (listed in the following table):

Challenge	Chapters
Inventory discrepancies across warehouses	2, 3, 4
Real-time inventory updates	5
Fraud detection	7
Cloud migration with zero downtime	9

Table 5: *Fusion-Mart business challenges mapped to solution chapters*

Each chapter uses Fusion-Mart scenarios to demonstrate practical patterns you can apply to your own platform.

CHAPTER 1
Evolution of Data Architecture

Introduction

The journey of data architecture is paved with both innovation and the lessons learned from past failures, often leading to complex, unmanageable, and unstable systems. This chapter dissects the evolution of data architectures, examining the critical shortcomings of historical approaches to justify the necessity of a structured governance framework. We will uncover why explicit governance is not merely a bureaucratic overhead, but a fundamental requirement for achieving a delicate balance between agility and stability in modern data environments. This historical analysis provides the crucial context for understanding the core requirements that motivate the Four-Gate Governance Model.

A deeper pattern underlies these failures: changes were implicitly trusted, or scrutinized, based on who proposed them rather than whether the changes themselves were safe. A schema modification from a senior engineer was trusted implicitly, while the same change from a junior developer faced scrutiny, even if the change itself was identical. This human-centric model becomes untenable as AI systems begin generating code, configurations, and data transformations. The Four-Gate Governance Model, introduced in this book, addresses this by shifting the question from *who wrote this?* to *is this change safe?*, a shift that becomes essential as AI-generated changes become commonplace.

As you work through each chapter, consider how these patterns apply to your own organization's data challenges.

Structure

This chapter covers the following topics:

- Genesis of data warehousing
- Rise of data lakes
- Lakehouse for unified data storage
- Industry adaptations and lessons learned

Objectives

By the end of this chapter, you will be able to: follow the progression from data warehouses to data lakes and finally to lakehouses, identifying the catalysts for each shift and the persistent challenges that remained. Evaluate the shortcomings of previous architectures, specifically focusing on ETL congestion and data integrity lapses that led to expensive corrections and regulatory breaches. Defend the necessity of formal governance by demonstrating how unregulated systems trigger systemic failures and build up significant technical debt. Finally, establish boundaries, reversibility, and evidence as the three foundational pillars that drive the implementation of the Four-Gate Governance Model.

Genesis of data warehousing

The evolution of data management reflects a continuous effort to align technological capabilities with escalating business demands. To understand the necessity of modern, balanced data frameworks, one must first examine the historical architectures that preceded them. Initially, organizations operated with fragmented data spread across siloed transactional systems, making comprehensive analysis nearly impossible. The data warehouse emerged as a solution, centralizing information to create a single source of truth for business intelligence. However, this approach introduced a fundamental tension between rigid governance and the need for analytical agility, leading to critical failures that directly inform the principles of contemporary data strategy.

Key considerations

Understanding the trajectory of data architectures requires acknowledging several pivotal points that highlight the shift from centralized control to balanced, modern frameworks. These considerations explain why past solutions became insufficient for the demands of modern enterprises:

- **The rise and limitations of centralization**: Data warehouses successfully solved the problem of data fragmentation by integrating disparate sources into a unified, reliable repository. This centralization enabled consistent reporting and strategic analysis that was previously unattainable. However, the strict governance required to maintain this integrity made the warehouse slow to adapt to new business questions and data sources, creating a significant bottleneck for innovation.
- **The inevitable governance-agility conflict**: The core challenge of the data warehouse era was the conflict between rigid governance and the business demand for agility. To ensure data quality and consistency, IT teams enforced rigorous data modeling and **extract, transform, load (ETL)** processes. This necessary control clashed with business users' need for rapid access to new data, leading them to create their own uncontrolled data silos and undermining the very concept of a single source of truth.

- **Learning from architectural failures**: The shortcomings of both extreme rigidity and unbounded fragmentation provide the foundational justification for modern data principles. The slow, inflexible data warehouse and the untrustworthy, chaotic data swamp revealed the limits of both approaches. These experiences highlight the value of architectures that build in clear data boundaries and auditable evidence from the start, allowing teams to move quickly without sacrificing the traceability that compliance and debugging require.

Therefore, historical data architectures reveal a persistent conflict between governance and agility, where the failures of rigid warehouses and chaotic data swamps motivate the core principles of modern, balanced systems.

Process or steps

The historical cycle of architectural failure followed a predictable pattern of problem, solution, and new problem. This process demonstrates how solving one challenge without a balanced approach often creates another:

1. **Cause**: Data fragmentation:

   Initially, enterprise data was trapped in operational silos like **customer relationship management** (**CRM**) or inventory systems. This fragmentation made it impossible to get a unified view of the business, causing inconsistent reporting and hindering strategic decision-making.

2. **Action**: Centralized warehousing:

   To solve fragmentation, organizations implemented data warehouses. They established rigorous ETL pipelines to extract data from operational systems, transform it into a consistent format, and load it into a central repository, transform it into a consistent format based on a predefined schema, and load it into a central repository, creating a trusted source for analytics.

3. **Result**: Rigidity and data swamps:

   The strict governance of the data warehouse became a bottleneck, slowing down the delivery of new insights. In response, business teams bypassed IT, creating ungoverned data repositories to achieve agility. This action led to data swamps, reintroducing data inconsistency, compliance risks, and a general lack of trust, even as teams tried to accelerate their work with a documented rollback plan for any failed data pipeline deployments.

Therefore, the process of solving data fragmentation with rigid centralization inadvertently created bottlenecks that drove users to create ungoverned data swamps, ultimately trading one set of problems for another.

Concrete example

The journey of a fictional retailer, Fusion-Mart, illustrates the practical consequences of this architectural evolution. Initially, Fusion-Mart struggled with disconnected data. Its marketing team could not measure the impact of online campaigns on in-store sales because the e-commerce and **point-of-sale** (**POS**) systems were separate. This lack of a unified view made it difficult to calculate customer lifetime value or optimize the supply chain effectively.

To address this, Fusion-Mart invested heavily in a corporate data warehouse. The project was a success; for the first time, the company could analyze holistic customer behavior, leading to more effective promotions and streamlined inventory management. However, this success created new demands. The marketing team wanted to quickly incorporate social media sentiment data to react to market trends in near real time. The data warehouse team, bound by its strict governance model, quoted a six-month timeline to properly model and integrate this new, unstructured data source.

Frustrated by the delay, the marketing analytics team used its own budget to dump raw social media data into an unmanaged cloud storage bucket. While they could now perform rapid sentiment analysis, their reports on brand perception directly contradicted sales figures from the official data warehouse. This discrepancy caused widespread confusion in executive meetings, eroded trust in both the marketing team's insights and the official corporate data, and made it possible to innovate on new data sources without compromising the integrity of financial reporting, which was protected by a feature flag isolating the experimental dashboard.

Therefore, Fusion-Mart's experience shows how a data warehouse can provide initial value, but its inherent rigidity can push business teams toward creating ungoverned data swamps, ultimately leading to conflicting reports and a loss of trust.

Cautions and limits

Reflecting on pre-lakehouse architectures reveals two critical risks that organizations must manage:

- **Over-correction risk**: Implementing excessively rigid governance in response to past ungoverned conditions stifles innovation and makes the data platform unresponsive to business needs.
- **Under-governance risk**: Pursuing agility without foundational controls leads back to the data swamp, poor data quality, security vulnerabilities, compliance failures, and eroded user trust.

Neither extreme is a viable long-term strategy. A successful modern architecture balances distributed autonomy with global consistency, embedding automated policy checks directly into deployment pipelines so teams can move quickly while maintaining compliance.

This book addresses these risks through the Four-Gate Governance Model:

- **Boundary gate**: Defines what changes are allowed within project scope, budget, and policy constraints, see *Chapter 2, Understanding Data Mesh, Lakehouse, and Medallion*
- **RB-30 gate**: Ensures any change can be rolled back within 30 minutes, see *Chapter 5, Building Scalable Data Pipelines*
- **Evidence gate**: Maintains audit trails for compliance and debugging, see *Chapter 6, Data Governance and Compliance*
- **AI proposal gate**: Evaluates AI-generated changes with the same rigor as human-authored ones, see *Chapter 7, MLOps for AI Model Deployment and Monitoring*

Each gate produces a clear decision: Go (proceed), Retry (fix and resubmit), or Kill (reject permanently).

Therefore, organizations must avoid the twin pitfalls of stifling rigidity and ungoverned fragmentation by building platforms that balance agility with essential controls for quality, security, and compliance.

Glossary

To ensure clarity, the key terminology discussed in this section is defined in the following:

- **Data warehouse**: A type of data management system designed to enable and support business intelligence activities, especially analytics. Data warehouses are intended to perform queries and analysis and often contain large amounts of historical data.

- **ETL**: A data integration process that combines data from multiple data sources into a single, consistent data store, which is loaded into a data warehouse or other target system.

- **Data swamp**: A deteriorated and unmanaged data lake that is either inaccessible to its intended users or is providing little value due to a lack of governance, metadata, and data quality.

- **Data lineage**: The lifecycle of data, including its origins and its journey and transformations over time. It provides a record of all data processing, which is essential for traceability, debugging, and compliance.

Expert notes

For experienced practitioners, the historical tension between governance and agility is best understood through the lens of architectural patterns like schema-on-write versus schema-on-read. Data warehouses enforced a strict schema-on-write model, where the data structure is defined before data is loaded. This ensures high quality and consistency, but at the cost of flexibility. In contrast, data lakes championed a schema-on-read model, allowing raw data to be loaded without a predefined structure, offering maximum flexibility but often leading to the data swamp issues discussed.

Modern architectures do not simply choose one over the other; they seek a synthesis. This evolution is summarized in *Table 1.1*. The principles of clear boundaries and auditable evidence are precursors to contemporary concepts like data mesh, where decentralized domain teams own their data products. This approach allows teams to manage their own data pipelines and schemas while adhering to global standards enforced through a shared platform. This allows them to develop new data products independently and still ensure interoperability, with changes to shared data models managed via a formal pull request that requires approval from stakeholders.

Therefore, the historical conflict between schema-on-write (warehouse) and schema-on-read (lake) models has evolved toward hybrid approaches that support decentralized ownership within a governed framework, a key tenet of modern data architecture.

The following table lists the technological evolution of data platforms:

Era	Primary focus	Key technology	Governance model	Failure mode
Data warehousing (1990s)	Centralized BI reporting	RDBMS (e.g., Oracle)	Top-down, IT-led	Inflexible; slow to adapt
Data lakes (2010s)	Unstructured Data and ML	HDFS, Spark	Decentralized, schema-on-read	Data swamp; poor quality
Lakehouse (2020s)	Unified analytics and AI	Delta Lake, Iceberg	Federated, programmatic	Governance complexity

***Table 1.1**: A summary of the technological evolution of data platforms*

Related topics

This section focuses on the historical justification for modern data governance. For related topics addressed elsewhere in this book:

- Lakehouse architecture implementation, see *Chapter 2, Understanding Data Mesh, Lakehouse, and Medallion*
- Data mesh patterns, see *Chapter 2, Understanding Data Mesh, Lakehouse, and Medallion*
- Medallion architecture design, see *Chapter 4, Medallion Architecture in Multi-cloud*

Rise of data lakes

The rise of data lakes was a direct response to the limitations of traditional data warehouses in handling the sheer volume and variety of modern data. These new platforms promised unprecedented agility, allowing organizations to store vast quantities of structured, semi-structured, and unstructured data in its raw format. This schema-on-read approach enabled faster data ingestion and flexible analysis for big data and machine learning. However, this flexibility often came at the cost of control, as early implementations frequently lacked the necessary governance to manage the data. This oversight showed that while rapid data collection is valuable, it must be paired with robust management, verified by quality assurance artifacts, to transform raw data into a trustworthy asset rather than a liability.

Key considerations

To understand why explicit governance is essential, it is important to examine the core issues that plagued early data architectures. The following points highlight the key challenges and requirements that emerged from these experiences:

- **Flexibility vs. reliability**: Data lakes were designed to overcome the rigid, predefined schemas of data warehouses, enabling teams to ingest diverse data types quickly. This schema-on-read capability accelerated data science and exploratory analytics, but without strong metadata management (data that describes other data) and quality controls, the data's reliability quickly eroded. Organizations can pursue agile data exploration and still maintain control through mechanisms like versioned access control policies.

- **Technical and organizational limitations**: The lack of governance created both technical and organizational problems. Technically, it led to poor data discoverability, inconsistent quality, and security vulnerabilities. Organizationally, it resulted in ambiguous data ownership and inconsistent practices, making it difficult to balance data accessibility with security and compliance mandates.

- **The need for foundational controls**: These widespread failures revealed a set of core requirements for building trustworthy data platforms. These include establishing clear boundaries for data assets, ensuring operational reversibility to correct errors, and maintaining auditable evidence of all changes. A simple feature flag provides a concrete way to enable or disable a data pipeline, offering a clear rollback path if issues arise.

Therefore, early data lakes prioritized flexibility over control, leading to technical and organizational failures that underscored the need for foundational governance principles like clear boundaries, reversibility, and auditable evidence.

Process or steps

The transformation of a promising data lake into an unusable data swamp typically follows a predictable sequence of events:

1. **Uncontrolled ingestion**: Driven by the goal of capturing all potentially useful data, teams begin loading raw data from various sources into the lake without standardized formats, quality checks, or comprehensive metadata. The cause is the desire for speed, leading to the action of bypassing upfront data curation.

2. **Knowledge erosion**: As more data accumulates without proper documentation, institutional knowledge about the data's origin, meaning, and quality resides only with the individuals who ingested it. When team members leave or change roles, this critical context is lost, and the result is a growing collection of datasets that no one trusts or understands.

3. **Loss of trust and utility**: Analysts and data scientists attempting to use the data encounter inconsistencies, errors, and duplication. This unreliability forces them to spend most of their time cleaning data rather than analyzing it, and eventually, they abandon the platform altogether. Analysts can gain speed by querying raw data directly, without sacrificing reliability when changes are tracked in an immutable commit log.

Therefore, a data lake becomes a data swamp through a process of uncontrolled data ingestion, which causes knowledge about the data to erode, ultimately resulting in a loss of trust and utility.

Concrete example

A national retail company launched a data lake initiative to create a single source of truth for customer behavior, combining online sales data, in-store transaction logs, and social media sentiment. Initially, the marketing team was able to quickly ingest this data and generate novel insights. However, without a governance framework, different departments uploaded data using inconsistent schemas. The **`customer_id`** from the e-commerce platform did not match the **`shopper_id`** from the in-store system, and date formats varied widely.

Within six months, the data lake was filled with conflicting and poorly documented data. A campaign designed to target high-value customers failed because the underlying data was unreliable, leading to significant financial loss. The compliance team could not verify data lineage for regulatory audits, putting the company at risk. The marketing team could launch campaigns faster, even as the compliance team verified data usage against a centrally managed audit trail link. This failure forced the company to halt new analytics projects and invest in a governance program to clean up the swamp, define data ownership, and implement quality controls.

Therefore, the retail company's ungoverned data lake quickly became unusable due to inconsistent identifiers and formats, causing a failed marketing campaign and highlighting the need for auditable data lineage and quality controls.

Cautions and limits

Implementing governance in a data lake environment is not without challenges. A primary risk is imposing overly restrictive controls that stifle the very agility the data lake was meant to provide. If governance processes are too bureaucratic, teams may revert to creating their own unmanaged data silos. Furthermore, achieving consensus on data ownership and standards across a large organization requires significant cultural change and executive sponsorship.

Another limitation is the technical complexity of available governance tools, which can require specialized skills to implement and maintain. To mitigate these risks, organizations should adopt an incremental approach, starting with foundational controls for the most critical data assets. It is crucial to automate governance checks within data pipelines wherever possible. Teams can innovate freely on new datasets while a feature flag allows for immediate rollback of a problematic data pipeline, ensuring that experiments do not jeopardize production systems.

Therefore, the primary challenges in data lake governance are balancing control with agility and managing organizational change; these can be mitigated by adopting an incremental, automated approach focused on critical data assets.

Glossary

Defining terms clearly enables broad understanding and still provides the precision needed for technical implementation, often captured in a shared **`glossary.md`** file:

- **Data lake**: A centralized repository that allows you to store all your structured and unstructured data at any scale. Unlike a data warehouse, it stores data in its raw format.
- **Schema-on-read**: An approach where a data structure or schema is applied to data as it is read, rather than when it is stored. This provides flexibility for handling diverse data types.
- **Metadata management**: The administration of data that describes other data. This includes information about data origin, format, structure, and usage, which is critical for discoverability and trust.

Expert notes

Experienced practitioners recognize that the failures of early data lakes were a key driver in the evolution of data platforms, as summarized in *Table 1.1*. This led to the development of the lakehouse architecture, which combines the flexibility of a data lake with the reliability and performance features of a data warehouse. A core principle in these modern systems is the use of open table formats that support **atomicity, consistency, isolation, durability** (**ACID**) transactions directly on data lake storage.

This technical foundation enables critical governance capabilities. For instance, every change to a dataset is recorded as a new version, creating an immutable history that supports error correction and rollbacks. Furthermore, this architecture facilitates the creation of verifiable evidence, where a single, permanent URL can point to a precise, immutable snapshot of the data, its schema, and its lineage. Advanced platforms can automate data processing without losing accountability, as every transformation is linked to a unique, verifiable build identifier, ensuring a complete and trustworthy audit trail.

Therefore, the evolution toward lakehouse architectures addresses historical data lake failures by integrating transactional capabilities and immutable versioning, which provides a technical foundation for robust governance and verifiable evidence.

Related topics

This section focuses on conceptual governance problems and the requirements that emerged from historical data architectures. For related topics addressed elsewhere in this book:

- Data mesh architecture, see *Chapter 2, Understanding Data Mesh, Lakehouse, and Medallion*
- Data governance implementation, see *Chapter 6, Data Governance and Compliance*

Lakehouse for unified data storage

The evolution of data platforms reflects a persistent effort to balance analytical agility with operational stability. Early **data warehouses** (**DWH**) provided strong governance but were rigid and expensive, while subsequent data lakes offered flexibility at the cost of reliability, often devolving into unusable data swamps. The lakehouse architecture arises from these historical lessons, creating a unified system that supports rapid innovation while providing the verifiable data quality and history essential for enterprise use. This modern approach is founded on core principles derived directly from the failures of past systems, as summarized in *Table 1.1*. A commit log that records every change to a dataset serves as the foundational proof of this stability.

Key considerations

Learning from historical architectural weaknesses reveals three essential requirements for building a resilient and agile data platform. These considerations form the foundation for a system that can evolve without collapsing under its own complexity:

- **Establish clear data boundaries**: The rigidity of traditional data warehouses demonstrated the need for well-defined yet adaptable data domains. By establishing clear boundaries through explicit schema contracts and ownership, organizations can empower teams to develop data products independently and still maintain system-wide coherence. A version-controlled boundary contract that defines schema and ownership for a data domain is a practical artifact for enforcing this principle.

- **Mandate comprehensive audit evidence**: The uncontrolled nature of early data lakes highlighted the critical need for traceability. A modern platform must capture complete data lineage, providing a transparent record of all changes, transformations, and access patterns. This ensures that all data is verifiable and trustworthy, enabling teams to build analytics with confidence while satisfying compliance requirements through artifacts like an access-controlled audit trail link.

- **Ensure rapid reversibility**: The operational difficulty of recovering from data corruption in complex systems drove the need for reliable restoration capabilities. Architectures must be designed for fast and predictable recovery, allowing operations to revert to a known-good state within a defined time window. This is achieved through immutable data versioning, which provides a concrete mechanism, like a single command to roll back a dataset to a previous, stable version.

Therefore, a successful data architecture requires explicit data boundaries, complete audit evidence, and rapid reversibility to prevent the failures of past DWH and data lake implementations.

Process or steps

Deriving core architectural principles from historical system failures follows a clear, three-step evaluation:

1. **Identify failure mode**: The process begins by analyzing a historical data architecture, such as a first-generation data lake, to pinpoint a specific cause of failure. This often involves untrustworthy data resulting from a lack of governance, leading to an inability to produce reliable analytics.

2. **Define corrective principle**: Based on the identified failure, an abstract guiding principle is formulated to prevent its recurrence. For instance, to combat untrustworthy data, the organization establishes a principle that all data transformations must be transparent, reproducible, and verifiable.

3. **Implement assurance mechanism**: The principle is then translated into a concrete technical control or process. To ensure verifiability, the platform might require that all data pipelines produce an immutable transaction log, allowing teams to innovate on data processing without sacrificing the integrity of the underlying datasets. A post-mortem document often links a data outage incident to the creation of a new data quality validation rule.

Therefore, by identifying historical failure modes, defining corrective principles, and implementing specific assurance mechanisms, organizations can systematically improve their data architecture.

Concrete example

An e-commerce company's first-generation data lake illustrates this process perfectly. The company initially allowed teams to ingest raw operational data into cloud storage with minimal oversight to accelerate the creation of analytics dashboards. The analytics team could build new dashboards with speed, even as new data sources were rapidly onboarded. However, this approach soon led to significant problems when two different marketing reports showed conflicting customer acquisition costs, creating confusion and eroding trust in the data team. A link in the data catalog that points directly to the source code commit hash for the transformation job that produced the customer acquisition cost table provides the necessary transparency.

The root cause was a classic data swamp failure: multiple teams had created their own conflicting data cleaning and enrichment pipelines with no shared standards or lineage. To correct this, the company established a principle of *verifiable data transformations*. They implemented a structured storage layer using an open table format like Apache Iceberg over their raw files. This new layer enforced schema validation on write and automatically versioned every change to a dataset, creating an immutable audit trail. Analysts could now trace the exact lineage of any data point, compare versions of a dataset, and confidently build reports based on a single, trusted source of truth.

Therefore, a real-world data swamp, caused by a lack of governance, was resolved by implementing a verifiable transformation principle enforced through a versioned and structured data storage layer.

Cautions and limits

While these principles are foundational, their implementation requires careful management to avoid common pitfalls. Organizations must introduce these controls incrementally, ensuring they enhance developer productivity without creating unnecessary friction. A primary risk is over-engineering the governance process, where rigid rules and manual reviews slow down development and negate the goal of agility. This can be mitigated by automating evidence collection and using declarative policy-as-code frameworks. Another challenge is cultural resistance from teams accustomed to operating with complete autonomy; this requires clear communication that frames these principles as enablers of quality and speed, not as bureaucratic hurdles. Finally, the underlying technology for versioning and lineage can be complex, so leveraging managed services and open standards is crucial to reducing the operational burden. A feature flag that allows a new data quality check to be tested in a non-blocking warning mode before being enforced is a key tool for managing this transition.

Therefore, implementing architectural principles requires mitigating the risks of over-engineering, cultural resistance, and technical complexity through automation, clear communication, and managed tooling.

Glossary

The following glossary defines the key technical terms introduced in this section:

- **Lakehouse architecture**: A modern data management architecture that combines the flexibility and cost-efficiency of data lakes with the data management and transactional features of data warehouses.
- **Schema enforcement**: A data governance feature that ensures data written to a table conforms to a predefined structure, preventing data quality issues.

Expert notes

Experienced practitioners can deepen these foundational principles with more advanced techniques. Advanced teams leverage declarative contracts to automate governance, allowing them to scale their data platform while maintaining strict compliance. For example, instead of writing imperative validation scripts, teams can define data quality rules in a version-controlled YAML file, which is easier to audit and manage. Furthermore, open table formats like Apache Iceberg, Apache Hudi, and Delta Lake are the critical technical enablers for these principles, providing transactional guarantees and schema evolution capabilities directly on object storage. Finally, it is essential to connect these technical mechanisms to business value; auditability directly supports regulatory requirements like the **General Data Protection Regulation** (**GDPR**), and rapid recovery is a prerequisite for meeting business-critical **service level objectives** (**SLOs**). A **continuous integration/continuous deployment** (**CI/CD**) pipeline that automatically validates a proposed schema change against downstream consumer compatibility tests before it can be merged is an example of such an advanced, automated control.

Therefore, experts enhance foundational principles by using declarative governance, leveraging open table formats, and explicitly linking technical capabilities like auditability and recovery to business objectives.

Related topics

This section focuses on foundational governance principles derived from historical architectures. For related topics addressed elsewhere in this book:

- Lakehouse architecture and Delta Lake, see *Chapter 2, Understanding Data Mesh, Lakehouse, and Medallion*
- Data mesh and decentralized ownership, see *Chapter 2, Understanding Data Mesh, Lakehouse, and Medallion*
- Data pipeline implementation, see *Chapter 5, Building Scalable Data Pipelines*

Industry adaptations and lessons learned

Understanding the historical context of how different industries adapted to evolving data challenges is paramount for making informed architectural decisions today. The journey from rigid, siloed systems to flexible, integrated platforms reveals common pain points and successful strategies. These lessons provide an invaluable foundation for designing robust and future-proof data architectures. Modern platforms enable rapid innovation on diverse datasets while maintaining strict data integrity, which can be verified through an immutable commit log. This evolution underscores a fundamental need: architectures must support rapid change without sacrificing stability, governance, or the ability to produce verifiable evidence of their operations.

Key considerations

The limitations of previous-generation data architectures directly inform the requirements for modern systems. The following highlights the primary failure modes that drove the industry's evolution:

- **Inability to handle modern data scales**: Traditional DWs were designed for structured data from predictable, transactional sources. They struggled to manage the three V's of modern data: volume, variety, and velocity. The explosion of unstructured and semi-structured data from e-commerce sites, social media, and **Internet of Things** (**IoT**) devices overwhelmed their rigid, schema-on-write models, making it impossible to capture valuable new insights.
- **Lack of analytical agility**: Pre-lakehouse systems were brittle and slow to change. Modifying a schema or integrating a new data source often requires months of effort, hindering the business's ability to respond to market changes. This lack of agility made it impractical to develop advanced, real-time applications for personalization, fraud detection, or dynamic risk analysis, which require iterative development and flexible data exploration.
- **Pervasive data silos**: In sectors like healthcare and finance, data was often locked in disparate, proprietary systems, from **Electronic Health Records** (**EHRs**) to mainframe transaction logs. This fragmentation prevented the creation of a holistic view of a patient or customer, blocking critical initiatives. Integrating these diverse and sensitive datasets was a primary driver for adopting centralized platforms that could unify information while enforcing security.

Therefore, traditional data architectures fell short because they were too rigid for the scale and diversity of modern data, too slow to support agile analytics, and too fragmented to provide a unified business view.

Process or steps

The migration from legacy systems to modern data platforms follows a consistent, three-step process driven by the need to overcome existing limitations:

1. **Cause**: Business demands exceed system capabilities:

   The process begins when a critical business objective cannot be met by the existing data architecture. This is often an inability to process new data types (e.g., customer clickstreams) or meet performance requirements for real-time analysis (e.g., fraud detection), creating a clear impetus for change.

2. **Action**: Adopt a flexible and scalable architecture:

   In response, the organization invests in a modern data platform, such as a data lake or lakehouse. This action involves establishing infrastructure capable of ingesting and storing raw data of any structure, decoupling storage from compute to allow for flexible processing and analysis. This process allows for the rapid development of new analytical models even as it enforces strict data access policies, which are defined and enforced through centrally managed identity provider groups.

3. **Result**: Enable advanced analytics with robust governance:

   With the new platform in place, the organization can build and deploy sophisticated applications that were previously impossible. This result delivers new business value while simultaneously providing the tools for essential governance, such as fine-grained access controls, data lineage tracking, and comprehensive audit logs, ensuring that innovation does not come at the cost of security or compliance.

Therefore, when business needs outpace technical capabilities, organizations adopt flexible architectures that unlock advanced analytics while providing the necessary governance controls.

Concrete example

A practical example of this process can be found in the **Banking, Financial Services, and Insurance** (**BFSI**) industry's approach to fraud detection. Legacy mainframe systems, while reliable for processing transactions, could not analyze patterns in real time to stop fraudulent activity as it occurred. The cause was clear: the rising sophistication of financial fraud required a more dynamic analytical capability. The bank's action was to implement a modern data platform that could stream transaction data in real time. This data was joined with historical customer profiles and known fraud patterns stored in the lakehouse. The result was the deployment of a machine learning model that could score transactions for fraud risk in milliseconds, allowing the bank to block suspicious payments before they were completed. Analysts can deploy new fraud detection algorithms quickly without disrupting core transaction processing, with each model's deployment and performance metrics captured in a version-controlled model registry. This move enabled a critical new business capability while maintaining the stability of the core transactional system.

Therefore, a bank facing sophisticated fraud moved from batch-oriented mainframes to a real-time streaming platform, enabling machine learning models to detect and block fraudulent transactions instantly.

Cautions and limits

Migrating to a modern data architecture introduces new challenges that require careful management to ensure success. While these platforms offer immense flexibility, they are not without risks. Organizations must proactively address potential pitfalls to balance innovation with stability. For instance, teams can empower more users with self-service analytics and still prevent data misuse by using attribute-based access control to dynamically filter sensitive data based on user roles. Key considerations include the risk of creating a data swamp if ingestion is not governed, the increased complexity of managing distributed systems, and the critical need for robust security controls to protect centralized data assets. Successful adoption depends on treating governance, security, and operational maturity as first-class citizens of the architecture, not as afterthoughts.

Therefore, adopting modern data platforms requires mitigating the risks of poor data quality, operational complexity, and security vulnerabilities through deliberate governance and controls.

Glossary

Refer to the following list for precise definitions of the key terms used in this discussion:

- **EHRs**: A digital version of a patient's paper chart. EHRs are real-time, patient-centered records that make information available instantly and securely to authorized users.
- **Health Insurance Portability and Accountability Act (HIPAA)**: The HIPAA of 1996, a US federal law that required the creation of national standards to protect sensitive patient health information from being disclosed without the patient's consent or knowledge.

Expert notes

The historical evolution of data platforms, summarized in *Table 1.1*, directly motivates the core requirements for a dependable data promotion system. The failures of past architectures reveal that progress and assurance must be built in, not bolted on. The inability of rigid data warehouses to adapt to new data formats underscores the need for explicit boundary rules, defined in version-controlled boundary contracts. The high-risk, monolithic change cycles of legacy systems highlight the necessity of robust rollback readiness, managed through mechanisms like feature flags. Finally, the stringent regulatory demands in industries like finance and healthcare demonstrate the non-negotiable requirement for comprehensive audit evidence, captured in an immutable log. This historical perspective shows that building a system for confident, rapid promotion requires embedding controls from the start, allowing developers to innovate freely within safe boundaries and still provide a complete audit trail.

Therefore, the architectural failures of the past directly inform the three pillars of a modern, reliable system: explicit data boundaries, safe rollback capabilities, and complete audit evidence.

Related topics

This section focuses on historical drivers and high-level architectural patterns across industries. For related topics addressed elsewhere in this book:

- Platform implementation patterns, see *Chapter 4, Medallion Architecture in Multi-cloud*

- MLOps and AI model deployment, see *Chapter 7, MLOps for AI Model Deployment and Monitoring*
- CI/CD for data engineering, see *Chapter 8, DevOps and CI/CD for Data Engineering*

Conclusion

This chapter illuminated the critical shortcomings of past data architectures, from monolithic data warehouses to unmanaged data lakes, demonstrating why explicit governance is indispensable for achieving both agility and stability. By understanding these historical failures, you now grasp the foundational requirements, boundaries, reversibility, and evidence that necessitate a robust framework like the Four-Gate Governance Model.

Having established the *why* behind modern data governance, our journey continues by exploring the leading architectural paradigms designed to address these challenges. The next chapter will explore these contemporary approaches, equipping you with the knowledge to evaluate and apply them effectively.

The principles of boundaries, reversibility, and evidence are not merely theoretical constructs; they are operational requirements that emerged from real-world system failures. Organizations that ignored these principles often faced costly data quality issues, compliance violations, and loss of stakeholder trust.

As you proceed through this book, you will see how these foundational requirements translate into concrete implementation patterns. The next chapter explores data mesh, lakehouse, and medallion architectures, showing how to draw your first boundary and make architectural decisions that support long-term governance goals.

Fusion-Mart's journey from disconnected data to governed architecture begins here, and throughout this book, you will see how they apply each principle to achieve measurable business outcomes.

CHAPTER 2
Understanding Data Mesh, Lakehouse, and Medallion

Introduction

Where does your project end and someone else's begin? This chapter answers the second question: *What are the boundaries?* by helping you define the non-negotiable limits at the boundary gate.

Modern data platforms are built on a few recurring architectural patterns, and understanding them clearly is the first step toward any successful data project. This chapter focuses on three such paradigms: data mesh, lakehouse, and medallion, and helps you see their strengths, limits, and trade-offs in practice. We will look at each of them through one concrete lens: how they behave when you face real-world constraints and need to draw a first boundary for your project. By the end of this chapter, you will be able to compare and contrast these three architectures and use them as lenses to think clearly about your project's scope, constraints, and non-negotiable limits at the boundary gate.

Structure

This chapter covers the following topics:

- Data mesh fundamentals
- Lakehouse architecture and Delta Lake
- Medallion architecture
- Data vault modeling
- Architecture decision matrix
- Lab: Boundary and RB-30

Objectives

By the end of this chapter, you will be able to compare data mesh, lakehouse, and medallion patterns using a decision matrix based on your organization's constraints rather than vendor marketing and explain when each pattern is the right choice and when it creates more problems than it solves for your specific situation. You will evaluate trade-offs between centralized and decentralized approaches, understanding how each affects autonomy and governance, while defining non-negotiable boundaries, including budget limits, security policies, and compliance requirements that no change can violate. Finally, you will implement the boundary gate to enforce these constraints automatically and prevent violations that compromise governance.

Data mesh fundamentals

Data mesh is a sociotechnical architecture that decentralizes data ownership to scale data-driven value creation. It shifts responsibility from a central data team to domain-oriented teams who generate and best understand the data. This approach addresses the bottlenecks and scalability issues of traditional, monolithic data platforms. By treating data as a product and providing a self-serve platform, organizations can accelerate innovation by empowering teams while ensuring enterprise-wide consistency through a small, version-controlled data contract that clearly states boundaries and responsibilities. This fosters agility and accountability, enabling faster delivery of high-quality data products.

Key considerations

Data mesh principles fundamentally reshape an organization's relationship with its data. The following points are central to its successful adoption:

- **Decentralized ownership**: The core principle is shifting data ownership to the business domains that are closest to the data. This decentralization assigns clear accountability, breaking down silos where a central team becomes a bottleneck. This approach empowers domains to move quickly and still maintain high data quality, verified through automated data quality checks embedded in the deployment pipeline.
- **Data as a product**: Domains must treat their data as a product they serve to internal customers. This requires a focus on making data discoverable, addressable, trustworthy, and self-describing. Adopting this product mindset ensures data is not just a technical byproduct but a valuable, reliable asset for the entire organization, with its integrity supported by reliability features such as atomic write operations and schema enforcement.
- **Federated governance**: To prevent chaos in a decentralized system, a federated governance model is essential. This involves a council of domain representatives and central platform owners who collaboratively define global standards, policies, and interoperability protocols. This allows for domain autonomy even as it enforces the non-negotiable limits required for security and compliance, managed through a central policy-as-code repository.

Therefore, successful data mesh adoption hinges on decentralizing ownership to business domains, treating data as a product with clear quality standards, and balancing autonomy with global rules through federated governance.

Process or steps

Implementing a data mesh architecture involves a structured transition from a centralized to a decentralized model. The process follows these key steps:

1. **Identify domains and ownership**: The initial cause of data bottlenecks is often unclear ownership. The first action is to map business capabilities to logical data domains, such as marketing or store operations. The result is that each domain gains clear accountability for its data assets, forming the foundation for decentralized management.

2. **Define data products**: Unreliable data often results from a lack of consumer focus. The action is for each domain to define its data products, complete with clear schemas, **service level agreements (SLAs)** defining freshness and quality expectations for batch or near real-time delivery, and documentation. This product-centric approach results in discoverable and trustworthy data that other domains can confidently consume, with schema enforcement providing a hard technical boundary. Note that this model applies most directly to batch and lakehouse workloads; for event streams, schema enforcement typically occurs at the producer or consumer level rather than as a centralized gate.

3. **Establish a self-serve platform**: Domain teams lacking data engineering skills can slow progress. The action is to build a central self-serve data platform that abstracts infrastructure complexity, providing tools for storage, processing, and sharing. The result is that domain teams can independently build, deploy, and manage their data products without relying on a central team, accelerating delivery while maintaining standards through a shared, versioned library of infrastructure-as-code modules.

4. **Implement federated governance**: Decentralization without standards leads to inconsistency. The action is to form a governance body that defines global rules for interoperability, security, and compliance. The result is a cohesive and secure ecosystem where data products from different domains can be easily combined, with compliance adherence tracked via an immutable audit log.

Therefore, the process involves identifying business domains to assign ownership, defining data as consumable products, enabling teams with a self-serve platform, and ensuring consistency through federated governance.

Concrete example

Consider a retail company, Fusion-Mart, adopting data mesh principles. The website analytics team, a business domain, owns all data generated by customer interactions on the company's e-commerce site. Previously, this data was sent to a central data lake, where the marketing team had to request access and transformations, causing significant delays.

Under the new model, the website analytics team uses the company's self-serve data platform to create a web traffic events data product. This product has a well-defined schema for events like **`page_view`** and **`add_to_cart`**, and its quality is guaranteed by the domain team. The data is discoverable through a central data catalog. The marketing team can now directly access this reliable data product via a standardized API to analyze campaign effectiveness in near real-time. This direct access accelerates marketing analysis without sacrificing governance, as access controls are embedded directly into the data product's interface and logged for security audits. This approach aligns with the decision-making framework for governance needs shown in *Figure 2.3*.

Therefore, a website analytics team owning its data and publishing it as a high-quality, discoverable web traffic events product allows a marketing team to consume it directly, accelerating analysis.

Cautions and limits

While data mesh offers significant advantages, it introduces challenges that require careful management. The shift to domain-oriented ownership is a significant cultural change that requires buy-in from leadership and retraining for teams accustomed to a centralized model. Without this cultural support, teams may resist taking on new responsibilities, undermining the entire initiative. A clear rollback plan, managed via feature flags, can de-risk the initial transition for pilot teams.

Furthermore, building a true self-serve data platform is a complex engineering effort. An under-resourced or poorly designed platform can create more friction than it removes, frustrating domain teams and hindering adoption. It is crucial to treat the platform as an internal product with its own dedicated team and roadmap. Finally, federated governance must strike a delicate balance; too much control stifles domain autonomy, while too little leads to a chaotic landscape of incompatible data products. This balance allows teams to innovate on their data products while still adhering to global standards, with compliance verified by automated policy checks in the CI/CD pipeline.

Therefore, organizations must manage the cultural shift to domain ownership, invest adequately in the self-serve platform, and carefully balance central governance with domain autonomy.

Glossary

The following are the key terms that have been discussed in this section:

- **Data mesh**: A sociotechnical architecture that decentralizes data analytics at scale by shifting ownership of data to business domains.
- **Domain-oriented ownership**: The principle that data should be owned and managed by the business domain teams that have the most context and expertise about it.

- **Data as a product**: The practice of treating data as a first-class product, with a focus on its quality, usability, and lifecycle management for its consumers.
- **Self-serve data platform**: A centralized platform that provides domain teams with the tools and infrastructure to build, deploy, and manage their own data products independently.
- **Federated computational governance**: A governance model where a central authority and domain representatives collaboratively create and automate global data policies, ensuring interoperability and security across the decentralized landscape.

Expert notes

For experienced practitioners, the power of data mesh lies in its application of product thinking and DevOps principles to the data domain. By treating data as a product, teams are incentivized to focus on consumer needs, leading to higher-quality and more valuable data assets. The self-serve platform is essentially an internal **platform as a service** (**PaaS**) for data, abstracting away low-level infrastructure and enabling rapid, repeatable deployments. This accelerates development even as it enforces architectural standards through versioned infrastructure templates.

Technically, this model thrives on automation. Governance policies should be expressed as code (policy-as-code) and automatically enforced by the platform. For instance, data quality checks, schema validation, and access control policies can be integrated into the CI/CD pipeline for data products. This ensures that every data product meets global standards before it is published. Advanced features like the ability to query data at a specific point in time become critical for auditability and reproducibility, providing a concrete mechanism for generating audit evidence.

Therefore, data mesh applies modern product and DevOps principles to data, relying on an automated, self-serve platform and policy-as-code to scale governance and quality.

Related topics

This section focuses on the conceptual framework of data mesh and its application to defining project boundaries. For related topics addressed elsewhere in this book:

- Self-serve data platform implementation, see *Chapter 5, Building Scalable Data Pipelines*
- Data governance and compliance see *Chapter 6, Data Governance and Compliance*

Lakehouse architecture and Delta Lake

The lakehouse paradigm represents a significant evolution in data architecture, creating a unified platform that merges the flexibility of data lakes with the reliability of data warehouses.

This approach addresses a core industry challenge: data lakes are cost-effective for storing vast, unstructured data but often lack the data quality and transactional controls needed for critical business intelligence. Conversely, data warehouses provide robust structure and performance but can be rigid and expensive. By leveraging open data formats on cloud storage and adding a transactional layer like Delta Lake, the lakehouse delivers **atomicity, consistency, isolation, durability** (**ACID**) guarantees, schema enforcement, and data versioning directly on the data lake. This is important because it enables organizations to build a single, reliable source of truth for analytics and machine learning without costly data duplication and complex pipelines.

Key considerations

Here are the most important points to consider when adopting a lakehouse architecture:

- The lakehouse model unifies the historically separate domains of data lakes and data warehouses. It allows organizations to store all their data, such as structured, semi-structured, and unstructured, in a single, low-cost location while still performing mission-critical analytics and business intelligence operations directly on that data. This significantly reduces the need for redundant systems and complex ETL processes, though some organizations may still maintain separate systems for workload isolation or specific performance requirements, streamlining the data pipeline and accelerating time-to-insight.

- Open table formats, including Delta Lake, Apache Iceberg, and Apache Hudi, are the foundational technologies that transform a standard data lake into a reliable lakehouse. While this chapter uses Delta Lake for concrete examples, the same principles apply to Iceberg and Hudi, and organizations should choose based on their ecosystem and requirements. It operates as an open-source storage layer that sits on top of existing cloud storage and introduces critical reliability features. These include ACID transactions for data integrity, schema enforcement to prevent data quality degradation, and time travel for auditability and error recovery.

- The features provided by Delta Lake are essential for establishing strong data governance and operational excellence. They enable teams to build robust, automated data pipelines that deliver high-quality data without sacrificing development speed. This progress is managed and still assured by clearly documenting the project's operational limits and responsibilities in a small, version-controlled boundary note or contract.

Therefore, the lakehouse, powered by Delta Lake, provides a unified and reliable data platform by adding transactional guarantees and governance features directly to cost-effective cloud data lakes.

Process or steps

Ensuring data reliability in a lakehouse follows a structured, three-stage process:

1. **Cause**: Ingest raw data:

   Data from diverse operational sources, which may have inconsistent schemas or contain errors, is ingested into the cloud data lake. In its raw form, this data poses a risk, as quality issues could corrupt downstream analytics and machine learning models if not properly managed.

2. **Action**: Apply transactional controls:

   When this data is written to a Delta table, the Delta Lake storage layer intercepts the operation to apply its core reliability features. The write is executed as an ACID transaction to ensure it either fully succeeds or fails. Schema enforcement validates the incoming data structure against the target table's schema, rejecting non-conforming records.

3. **Result**: Achieve a reliable state:

   The transaction commits successfully, and the new data is added to the table as a new, versioned state recorded in the transaction log. The result is clean, consistent, and auditable data ready for consumption, even as new data streams in. This entire process is governed by a feature flag that allows for immediate rollback of the pipeline configuration if an issue is detected. For batch and micro-batch workloads, this rollback can restore previous data states; for true streaming, rollback applies to configuration changes while data recovery relies on replaying from upstream sources.

Therefore, the process for ensuring reliability involves ingesting raw data, applying Delta Lake's transactional and schema controls during the write operation, and producing a consistent, versioned, and trustworthy dataset as the result.

Concrete example

A practical example of the lakehouse in action is an e-commerce company managing its product inventory and sales data. The company uses a `MERGE` operation to update its master inventory table with data from a stream of new sales, a process that must be completely reliable to avoid overselling products. The ACID transaction guarantees that the sales record and the inventory deduction are completed as a single atomic unit. Simultaneously, a data feed from a new supplier attempts to write records with an incorrect schema to the sales table; schema enforcement automatically rejects this write, preventing data corruption and maintaining the integrity of the system. When a faulty pricing update causes incorrect sales figures, the data team can use time travel to revert the table to its state just before the error, minimizing business impact without a lengthy data restoration process. This allows the business to move forward with new product launches while ensuring that a link to the audit trail for the rollback is automatically generated and stored.

Therefore, an e-commerce platform uses Delta Lake's ACID transactions for reliable inventory updates, schema enforcement to block bad supplier data, and time travel to instantly roll back a faulty pricing update.

Cautions and limits

While the lakehouse architecture offers powerful capabilities, organizations must be aware of its potential challenges and prerequisites. Adopting this model requires a shift in mindset from traditional database management, and teams may need training on concepts like copy-on-write and optimistic concurrency. Furthermore, features like time travel, which retain historical data versions, can lead to increased storage costs if not managed properly; implementing a data retention policy using the `VACUUM` command is essential for cost control.

Note that time travel and instant rollback capabilities are most effective for batch and micro-batch workloads. For true streaming scenarios with continuous data flow, rollback strategies require additional patterns, such as replaying from upstream Kafka topics or restoring from checkpoints, since the continuous nature of streaming limits point-in-time table restoration. *Chapter 5, Building Scalable Data Pipelines*, explores these streaming-specific recovery patterns in detail.

The most significant prerequisite is a commitment to upfront design and governance. Defining project scope, budget, and irreversible limits in a small, version-controlled boundary note or contract is not optional; it is a critical step to prevent uncontrolled complexity and ensure the architecture meets its objectives. This allows for rapid development and still provides a clear, auditable record of all architectural decisions and their constraints, with a documented manual override procedure for emergencies.

Therefore, a successful lakehouse adoption requires managing the learning curve, controlling storage costs from data versioning, and committing to strong upfront governance by defining project limits in a small, version-controlled boundary note or contract.

Glossary

Here are definitions for key terms used in this section:

- **Service level indicator (SLI)**: A quantitative measure of some aspect of the level of service that is provided, such as uptime or data freshness.
- **Service level objective (SLO)**: A target value or range for a service level, measured by an SLI, that a system aims to meet.
- **Service level agreement (SLA)**: A formal contract between a service provider and a customer that defines the expected level of service, often with penalties for non-compliance.
- **ACID transactions**: A set of properties of database transactions (ACID) intended to guarantee data validity despite errors, power failures, and other mishaps.

- **Schema enforcement**: A feature that prevents data with a different schema (e.g., wrong data types or new columns) from being written to a table, thereby protecting data quality.
- **Time travel**: The capability to access and query previous versions of a data table, enabling use cases like auditing, data rollback, and reproducing experiments.

Expert notes

For experienced practitioners, it is crucial to understand that Delta Lake's reliability features are all derived from its transaction log, stored in the **`_delta_log`** directory alongside the data files. This log is the single source of truth; every operation that modifies the table is recorded as an ordered, atomic commit. This design enables optimistic concurrency control, allowing multiple readers and writers to operate on a table simultaneously with minimal conflict, which is ideal for modern analytics workloads. This robust foundation is what makes it possible to implement structured architectural patterns like the medallion architecture, as depicted in *Figure 2.1*, where data is refined through bronze, silver, and gold tables with guaranteed quality at each stage. Advanced configurations and data quality rules can be codified and versioned alongside your data platform configuration and boundary contracts, ensuring that governance evolves with the data platform and still provides an immutable audit trail of all changes.

Therefore, Delta Lake's power stems from its transaction log (**`_delta_log`**), which enables optimistic concurrency and provides the reliable foundation required for sophisticated data refinement patterns like the medallion architecture.

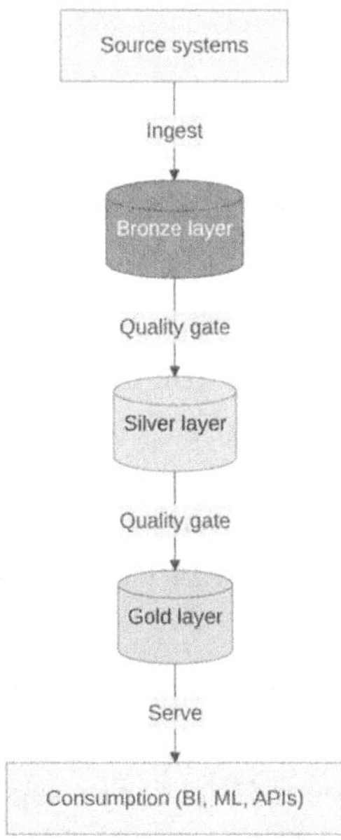

***Figure 2.1**: Data refinement process*

Related topics

This section focuses on foundational lakehouse concepts and Delta Lake capabilities. For related topics addressed elsewhere in this book:

- Multi-cloud data formats comparison, see *Chapter 4, Medallion Architecture in Multi-Cloud*
- Performance tuning and cost optimization, see *Chapter 5, Building Scalable Data Pipelines*
- Auditing, transaction logs, and rollback playbooks, see *Chapter 6, Data Governance and Compliance*
- Data vault 2.0 implementation patterns, see *Figure 2.2* for an overview

Medallion architecture

The medallion architecture provides a structured, multi-layered approach to building reliable data pipelines, transforming raw, often unreliable data into validated, high-value assets. This pattern is essential for modern data platforms because it establishes a clear, repeatable process for improving data quality and ensuring traceability from source to consumption. By progressively refining data through bronze, silver, and gold layers, organizations can systematically build trust in their data assets while maintaining a complete, auditable history of all transformations. This methodology helps make non-negotiable limits explicit and provides operational proof of a governed process through clear, versioned data lineage from source to consumption.

Key considerations

Implementing a medallion architecture requires careful thought around its core principles to maximize its benefits for governance and data quality. The following points are central to a successful deployment:

- The architecture's strength lies in its progressive data refinement across three distinct layers: bronze for raw data, silver for cleaned data, and gold for business-ready aggregates. This layered separation of concerns allows data engineering teams to isolate different stages of data processing, enabling them to improve data quality incrementally without impacting downstream consumers. It also lets teams introduce new gold tables safely, testing them with a small group of consumers before broad rollout.

- The bronze layer's immutability is a non-negotiable principle for ensuring auditability and traceability. By storing source data in its original, unaltered format, organizations create a permanent historical record that can be reprocessed to fix errors or accommodate new business logic. The immutability of the bronze layer is operationally enforced through write-once storage policies, providing a non-repudiable audit trail.

- This pattern inherently establishes clear boundaries and quality gates between each layer, making it a powerful tool for defining and enforcing data governance standards. This structure allows for rapid development of new data products in the gold layer without compromising the integrity of the foundational silver layer. The successful validation of a data quality test before promotion to the silver layer serves as a concrete assurance artifact for this governance.

Therefore, the medallion architecture's value comes from its disciplined, three-layer structure that enforces progressive data quality, guarantees auditability through an immutable raw layer, and establishes clear governance boundaries.

Process or steps

The flow of data through the medallion architecture follows a logical sequence of refinement, as shown in *Figure 2.1*:

1. **Ingest raw data**: The process begins when data from source systems is landed in the bronze layer. This action is a direct copy, preserving the original format, structure, and content without any cleaning or transformation, which ensures a complete and immutable historical archive is created for future reprocessing and auditing.
2. **Clean and conform**: Data is loaded from the bronze layer and subjected to a series of transformations to create the silver layer. This involves applying data quality rules, parsing complex types, enriching records with other data sources, and conforming datasets into a standardized, entity-based model (e.g., Customers, Products). The result is an enterprise-wide, single source of truth that provides a reliable foundation for all downstream analytics.
3. **Curate and aggregate**: Analysts and data scientists consume data from the silver layer to build curated datasets for the gold layer. This step involves creating project-specific aggregations, feature engineering for machine learning models, or building performance-optimized data marts. These gold tables provide high-value, consumption-ready data assets that directly power **business intelligence** (**BI**) dashboards and applications, even as new data sources are added.

A failed data quality check automatically triggers a notification and quarantines the problematic data, with a link to the failed pipeline run serving as the concrete evidence of enforcement.

Therefore, the process moves data from a raw, immutable bronze layer, through a cleaned and conformed silver layer, to a highly curated and aggregated gold layer ready for business consumption.

Concrete example

Consider Fusion-Mart, an e-commerce company processing online sales data. The medallion architecture provides a clear path to turn raw transaction events into actionable business insights.

In the bronze layer, raw, nested JSON events from the company's web servers are ingested into a table like **`raw_sales_events`**. Each record is a verbatim copy of the source data, complete with timestamps and all original fields, providing a perfect, auditable archive. This allows analysts to build new dashboards from the gold layer and still trace any metric back to its raw source event in the bronze layer for validation.

Next, a pipeline processes this raw data to populate the silver layer. It parses the JSON, corrects data types (e.g., converting string dates to timestamps), filters out test transactions, and joins customer information from a separate system. This creates clean, conforming tables such as **`Orders`** and **`Order_Items`**, which represent a trusted, master view of the core business entities.

Finally, the gold layer is built from the silver tables to serve specific business needs. A data analyst might create a **`monthly_regional_sales`** summary table that aggregates sales figures by product category and geographic region. This aggregated table is optimized for performance and directly powers a regional sales dashboard, providing fast, reliable metrics to business leaders. A direct hyperlink from a dashboard metric to the specific query and data version used for its calculation provides an immutable proof of its origin.

Therefore, an e-commerce company can use the medallion pattern to transform raw JSON clickstream data (bronze) into clean **`Orders`** tables (silver) and finally into a **`monthly_regional_sales`** dashboard (gold).

Cautions and limits

While the medallion architecture provides significant benefits, teams must be aware of its potential challenges to ensure a successful implementation. The primary risk is over-engineering; not every data pipeline requires the full three-layer treatment. For simple, low-impact data, a two-layer (bronze-to-silver) approach may be sufficient. Organizations can manage the increased processing latency inherent in a multi-layer system without sacrificing the benefits of data quality by clearly defining SLAs for each layer.

Another consideration is the potential for increased data latency, as each layer adds a processing step. For use cases requiring near-real-time data, this architecture might need to be augmented with streaming patterns. Finally, the separation of layers introduces governance overhead. Maintaining schemas, access controls, and data quality rules across multiple layers requires disciplined practices and automation, such as using policy-as-code to define and enforce contracts. A small, version-controlled boundary note or contract explicitly defines the acceptable latency for gold layer data, serving as a non-negotiable limit.

Therefore, be mindful of over-engineering, potential data latency, and the need for disciplined governance to avoid turning the architecture's structure into a liability.

Note on streaming workloads

The architectural patterns discussed in this chapter are primarily optimized for batch and micro-batch processing. When incorporating streaming data, consider the following adaptations:

- **Ingestion**: Streaming data typically lands in the bronze layer through message queues (such as Kafka topics) and is processed in micro-batches or near real-time windows.
- **Processing**: True real-time use cases may require dedicated streaming pipelines that operate alongside, rather than within, the traditional bronze-silver-gold flow.
- **Recovery**: Point-in-time rollback is limited for streaming; plan for replay-based recovery from source topics or checkpoint restoration.

These streaming-specific patterns are covered in depth in *Chapter 5, Building Scalable Data Pipelines*.

Glossary

This section discusses the following key terms:

- **Medallion architecture**: A data architecture pattern that organizes data into three distinct layers: bronze, silver, and gold, to progressively improve data quality and usability.
- **Data lineage**: A record that details the origin of data and traces its journey and transformations through various systems and processing steps.
- **Immutability**: A principle stating that data, once written to a storage layer, cannot be altered or deleted. This is critical for creating auditable, reproducible data records.
- **Conformed data**: Data that has been standardized, cleaned, and integrated from multiple sources to provide a single, consistent view of core business entities.

Expert notes

For advanced implementations, the medallion architecture serves as a foundational pattern that can be enhanced with other methodologies and technologies. For instance, a data vault model, as illustrated in *Figure 2.2*, can be implemented within the silver layer to provide highly auditable, fine-grained historical tracking of business relationships. This allows practitioners to gain deep auditability without disrupting the consumption patterns of the gold layer. The successful execution of an atomic `MERGE` operation, logged with its transaction ID, provides concrete proof of a reliable and auditable update.

Furthermore, the reliability of a medallion architecture is significantly bolstered by leveraging the capabilities of modern data lakehouse platforms. Features such as ACID transactions on lakehouse tables ensure that operations are completed fully or not at all, preventing data corruption. Schema enforcement and evolution prevent data quality degradation from upstream changes, while time travel capabilities allow for easy rollbacks and reproducible queries, providing a safety net for development and operations.

Therefore, experts can enhance the medallion architecture by integrating it with a data vault for audibility and by leveraging modern lakehouse features like ACID transactions and schema enforcement for operational resilience.

Related topics

This section focuses on medallion architecture principles and structure. For related topics addressed elsewhere in this book:

- Cloud platform implementation, see *Chapter 4, Medallion Architecture in Multi-Cloud*
- Data pipeline orchestration, see *Chapter 5, Building Scalable Data Pipelines*
- Data mesh comparison, see *Table 2.2* in this chapter

Data vault modeling

Data vault is a data modeling technique designed for organizations that need a long-term, auditable view of their core business data. In environments where customer, sales, marketing, and service information is scattered across many operational systems, it offers a consistent way to integrate that data without losing history. By separating stable business keys, the relationships between them, and their changing descriptive attributes, data vault creates a flexible, historically complete store that supports regulatory compliance, explainable analytics and AI, and ongoing business change. This is because every change is stored as a new, time-stamped row; teams can always reconstruct what the data looked like at any point in time and where it came from.

Key considerations

When evaluating a data vault, organizations must weigh its powerful capabilities against its structural demands. The following points are central to this consideration:

- **Auditability and lineage by design**: Data vault's fundamental structure is built to capture a complete, immutable history of all data. By separating business keys from their descriptive attributes and tracking every change over time, it provides an inherent audit trail. This is non-negotiable for AI accountability, allowing teams to reconstruct the exact state of data used for any past decision, even as they can trace data provenance through a metadata-driven lineage graph.

- **Integration and flexibility**: The methodology excels at integrating data from numerous, diverse source systems without requiring significant re-engineering of the core model. Its component-based design allows new data sources or attributes to be added gracefully, preventing the brittleness common in other modeling approaches. This ensures the data warehouse can evolve with the business and still maintain referential integrity, which is validated by a suite of automated data quality tests that run with each data load.

- **Complexity and operational overhead**: The benefits of data vault come with increased structural complexity. The model typically results in a higher number of tables and joins compared to dimensional models, which can present a steeper learning curve and require more sophisticated data ingestion processes. Organizations can proceed with this architecture without sacrificing query performance by implementing a dedicated feature flag and rollback plan for any changes to the data transformation logic.

Therefore, data vault provides unparalleled auditability and flexibility for integrating diverse data sources, but this comes at the cost of increased model complexity and operational overhead that must be managed.

Process or steps

Implementing a data vault model involves three key steps that systematically separate business logic from historical context:

1. **Establish core business entities (Hubs)**: The first step is to identify the core business concepts, such as Customer or Product, and load their unique business keys (e.g., `customer_id`, `product_sku`) into `Hub` tables. This action creates a stable, non-volatile list of all business entities, providing a reliable anchor for all related data. The result is a centralized, de-duplicated master list of business keys, with each entry's origin tracked via a record source attribute.

2. **Define relationships (Links)**: Next, you model the transactions or associations between the core business entities. This is achieved by creating `Link` tables that connect two or more `Hubs`. For example, a `Link_Customer_Order` table would store the hash keys from the Customer and Order hubs, capturing the fact that a specific customer placed a specific order. This action builds the structural relationships of the business process, resulting in an auditable record of how entities interacted over time.

3. **Store descriptive attributes (Satellites)**: Finally, all descriptive and contextual data about entities or their relationships is loaded into `Satellite` tables. Each `Satellite` is attached to a parent `Hub` or `Link` and stores attributes like `customer_name` or `order_status`. Any change to an attribute causes a new, time-stamped row to be inserted, as demonstrated in the pattern at *Listing 2.1*. This action preserves all historical versions of the data, resulting in a complete and traceable history of every attribute, which is essential for point-in-time analysis and AI explainability.

Therefore, the data vault process isolates stable business keys in `Hubs`, connects them with `Links` to model relationships, and tracks all historical changes to descriptive data in `Satellite`.

The following listing shows data vault 2.0 Hub/Link/Satellite with hashed keys, pattern for long-term history, and compliance:

```
-- models/hubs/hub_customer.sql
-- This Hub model captures the core business entity 'Customer'.
```

```
WITH source_data AS (
    SELECT
        customer_id,
        load_date,
        record_source
    FROM {{ ref('stg_customers') }}
)

SELECT
    MD5(customer_id) AS customer_hk,
    customer_id,
    load_date,
    record_source
FROM source_data;

-- LINK MODEL
-- models/links/link_customer_order.sql
-- This Link model establishes the relationship between a Customer and an
Order.

WITH source_data AS (
    SELECT
        customer_id,
        order_id,
        load_date,
        record_source
    FROM {{ ref('stg_orders') }}
)

SELECT
    MD5(CONCAT(customer_id, '|', order_id)) AS link_customer_order_hk,
    MD5(customer_id) AS customer_hk,
    MD5(order_id) AS order_hk,
    load_date,
    record_source
FROM source_data;

-- SATELLITE MODEL
-- models/sats/sat_customer_details.sql
-- This Satellite model stores descriptive, time-variant attributes for a
Customer.
```

```
WITH source_data AS (
    SELECT
        customer_id,
        customer_name,
        customer_address,
        load_date,
        record_source
    FROM {{ ref('stg_customers') }}
)

SELECT
    MD5(customer_id) AS customer_hk,
    customer_name,
    customer_address,
    MD5(CONCAT(customer_name, '|', customer_address)) AS hash_diff,
    load_date,
    record_source
FROM source_data;
```

***Listing 2.1**: Data vault 2.0 Hub/Link/Satellite*

Concrete example

To illustrate the data vault process, consider a Fusion-Mart customer, identified by **customer_id 'CUST-1138'**, who makes a purchase. Initially, the system ingests the customer's data. The business key **'CUST-1138'** is hashed and loaded into the **Hub_Customer** table, creating a permanent anchor for this entity. Simultaneously, descriptive details like the customer's name and address are loaded into a **Sat_Customer_Details** table, linked directly to the **Hub_Customer** record and time-stamped with the load date. When the customer places an order with **order_id 'ORD-909'**, that business key is loaded into the **Hub_Order** table. A **Link_Customer_Order** table is then populated with the hash keys from both **Hub_Customer** and **Hub_Order**, formally recording the transaction. This structure, visualized in *Figure 2.2*, allows the business to add new data, like shipping information in a new satellite, without altering the existing core model. Even as new data sources are integrated, the model remains stable and still provides a complete historical view, especially when combined with atomic, logged write operations on the underlying lakehouse or warehouse platform.

Therefore, a customer's identity is stored in a **Hub**, their changing details in a **Satellite**, and their purchase transaction in a **Link**, creating a flexible and historically accurate record.

Cautions and limits

While data vault provides a robust framework for auditability, its adoption requires careful consideration of its inherent complexities. The highly normalized structure results in a large number of tables, which can make ad-hoc querying difficult for business users accustomed to simpler dimensional models. To mitigate this, organizations must plan for a separate presentation layer (e.g., the gold layer in the medallion architecture, as seen in *Figure 2.1*) where data is denormalized for analytical use. Furthermore, the development of data loading pipelines can be more intricate, demanding standardized patterns and automation to manage effectively. Teams can manage this complexity without compromising data integrity by combining data vault with platform features such as schema enforcement and governed lakehouse storage, as introduced earlier in this chapter. Acknowledging the steeper learning curve for developers and analysts is also crucial; success depends on proper training and a commitment to the methodology's principles.

Therefore, the primary limitations of data vault are its query complexity and development overhead, which can be managed by building a dedicated analytics layer and using standardized automation patterns.

Glossary

The following are the key terms that have been discussed in this section:

- **Business key**: A natural identifier from a source system that is meaningful to the business, such as a `customer_id` or `product_sku`. It is the cornerstone of a `Hub` table.
- **Hub**: A core data vault component that contains a unique list of business keys for a single business entity (e.g., customers, products). It acts as the anchor of the model.
- **Link**: A data vault component that establishes a relationship or transaction between two or more hubs. It represents the associations within the business process.
- **Satellite**: A data vault component that stores descriptive, time-variant attributes for a parent `Hub` or `Link`. It tracks historical changes to data, providing a complete audit trail.
- **Hash key**: A system-generated unique identifier, created by applying a hash function to a business key or a combination of keys. It serves as the primary key in Hubs and Links to facilitate efficient, anonymized joins.

Expert notes

For experienced practitioners, maximizing the value of a data vault implementation involves integrating it within a broader data architecture and leveraging advanced platform features. Data vault often serves as an ideal historical and integrated silver layer within a medallion-style architecture, as depicted in *Figure 2.2*. This layer provides a fully auditable foundation from which curated, business-facing gold tables can be derived for analytics and AI modeling. To ensure data integrity during ingestion, all loading processes must be atomic. Using operations like the `MERGE` command on lakehouse or warehouse tables guarantees that data writes are all-or-nothing, preventing partial or corrupt updates. This allows for continuous data loading even as analytical queries are running, and still provides a mechanism for immediate rollback when combined with platform features such as time-travel or table versioning.

Therefore, advanced data vault usage involves positioning it as the auditable silver layer in a medallion architecture and enforcing data integrity with atomic write operations.

The following figure shows the data vault core model, such as **`Hub`**, **`Link`**, **`Satellite`**, and fit with medallion layers:

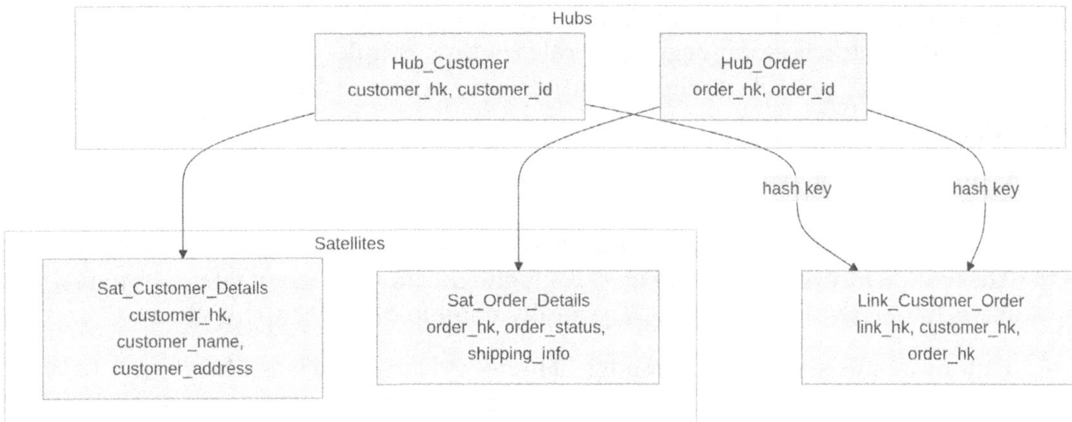

***Figure 2.2**: Data vault core model*

Related topics

This section focuses on data vault 2.0 foundational concepts. For related topics addressed elsewhere in this book:

- Architecture selection frameworks, see *Table 2.1* in this chapter
- Data pipeline implementation, see *Chapter 5, Building Scalable Data Pipelines*
- Performance tuning strategies, see *Chapter 10, Scaling Data Platforms with Optimization*

Architecture decision matrix

Selecting the right data architecture is a critical decision that balances business goals, technical requirements, and governance constraints. A well-chosen pattern accelerates development and delivers value, while a poor choice creates technical debt and operational friction. This section introduces a simple architecture decision matrix you can use to narrow down options such as data mesh, lakehouse/medallion, or medallion combined with data vault, and to record the final choice in a lightweight, version-controlled policy (for example, a boundary contract in YAML).

Key considerations

Before selecting an architecture, it is essential to evaluate several foundational aspects of the project. These considerations help frame the decision-making process and ensure the final choice is robust and sustainable:

- **Governance and compliance first**: Organizational policies, data ownership rules, and regulatory requirements must be the first filter. As shown in the governance-first flow in *Figure 2.3*, strict needs such as long-term, auditable retention can immediately narrow the architectural options. This allows you to stay flexible on technical choices while still enforcing non-negotiable rules, with a simple, version-controlled checklist (for example, *Table 2.3*) providing auditable proof of due diligence.

- **Workload characteristics**: The nature of the data workload is the primary driver for comparing candidate patterns. Factors such as data volume, ingestion velocity, query patterns, and performance SLAs determine whether you lean toward a batch-oriented cloud DWH + medallion, a streaming-heavy lakehouse, or a domain-oriented mesh. The workload-driven flow in *Figure 2.4* highlights how different needs, from real-time alerts to deep historical analysis, naturally point to different architectures.

- **Pattern nuances and hybrid models**: High-level labels such as *Medallion* or *Lakehouse* often hide important differences. In practice, many platforms combine patterns, for example, a cloud DWH + medallion for analytics, enhanced with a data vault layer where cross-domain auditability is critical. A concise comparison like *Table 2.2* helps clarify when a simple medallion approach is sufficient and when a hybrid medallion + data vault model is justified.

Therefore, prioritize governance and workload requirements to filter architectural options, then perform a detailed comparison of the remaining candidates to make an informed final decision.

The following figure shows a decision flow for selecting an architecture based on governance needs. Observe how strict compliance requirements, like long-term retention, significantly narrow the viable architectural patterns.

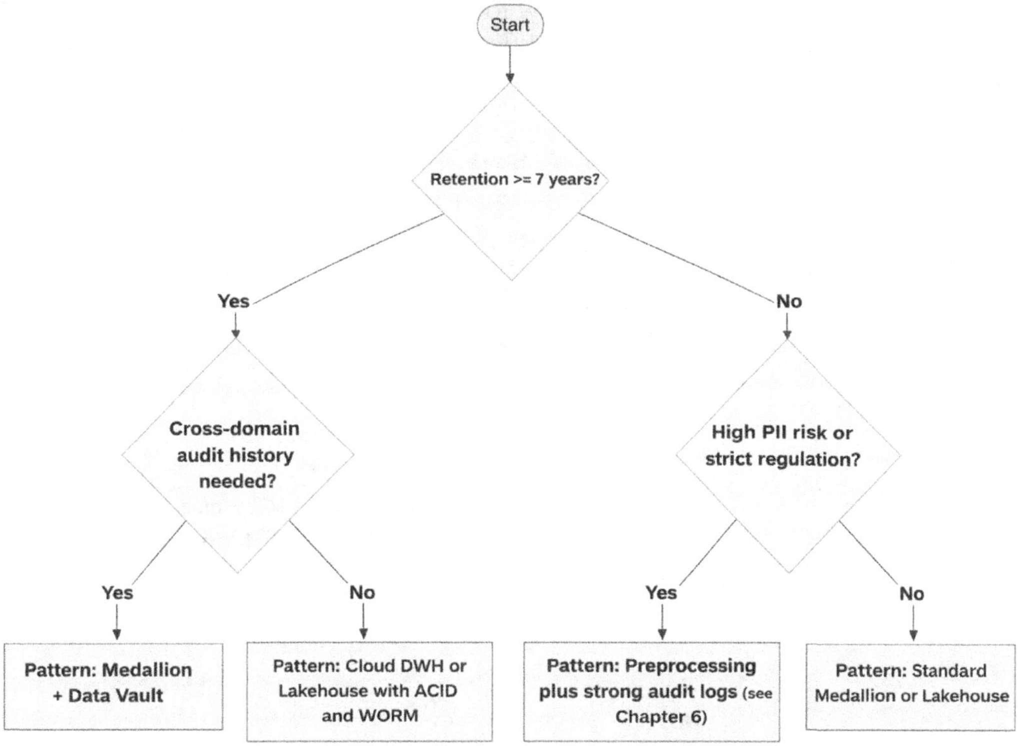

***Figure 2.3**: Decision flow for selecting an architecture based on governance needs*

The following table shows an architecture selection:

Axis	Options	Decision hint
Retention or regulation	<3 y / ≥7 y	≥7 y | ACID tables and WORM baseline
Update frequency	Batch or near real-time	Near-RT | streaming path
Data shape	Append-heavy or frequent updates/ deletes	Frequent updates | Delta/Iceberg/Hudi
Responsibility	Centralized or domain-distributed	Distributed | Mesh contracts and SLOs
Privacy	Low or high	High | pre-processing (mask/anonymize) and audit logs (see *Chapter 6, Data Governance and Compliance*)

***Table 2.1**: Architecture selection*

The following figure shows the decision flow for selecting an architecture based on workload characteristics. The key takeaway is that different business needs, like streaming SLAs or auditability, lead to fundamentally different architectural choices.

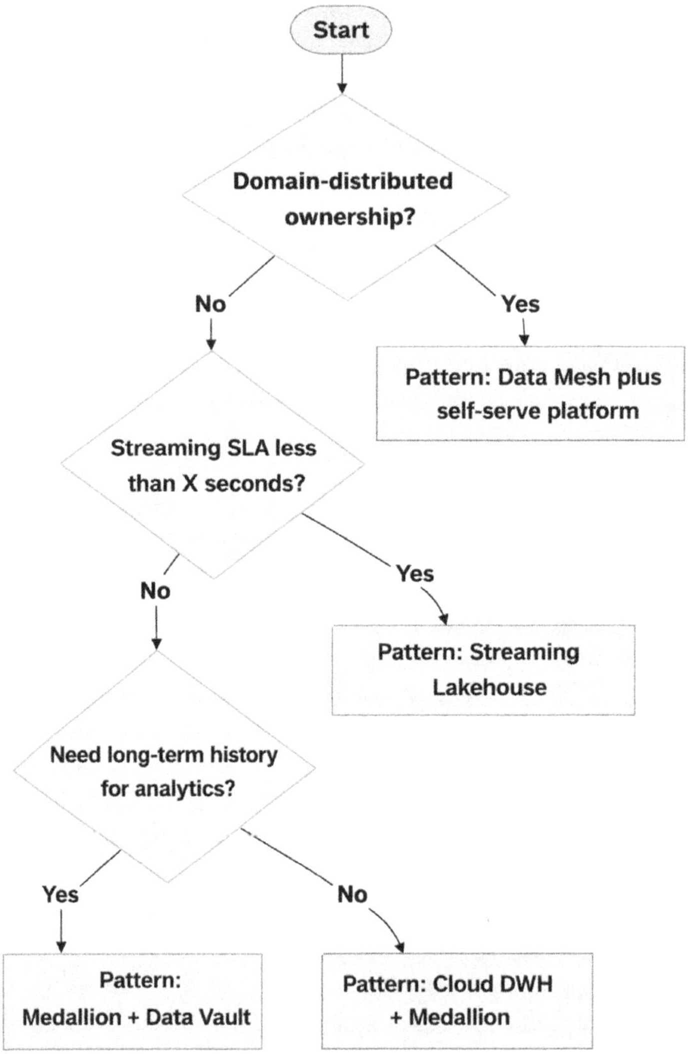

Figure 2.4: *Decision flow for selecting an architecture based on workload characteristics*

Process or steps

Selecting the right data architecture involves a systematic, multi-stage process that narrows options from general patterns to a specific, validated choice:

1. **Perform high-level scoping**: Use a broad decision matrix, such as *Table 2.1*, to identify a shortlist of suitable patterns. Align basic characteristics, batch vs. streaming, analytical vs. operational, with the strengths of each architecture.
2. **Apply governance-based filtering**: Filter the shortlist using your organization's non-negotiable governance and compliance requirements. Following *Figure 2.3*, remove any pattern that cannot satisfy retention, sovereignty, or auditability needs.
3. **Execute workload-driven selection**: Compare the remaining options against concrete workload requirements, as in *Figure 2.4*. Evaluate how each pattern handles ingestion rates, transformation complexity, and query SLAs, and converge on a single recommended architecture.
4. **Codify the decision and boundaries**: Record the chosen pattern and its key constraints as policy-as-code. For example, using a small YAML boundary contract that captures scope, budget, and irreversible limits. Keeping this file under version control turns architecture choice into a living, reviewable contract.

Therefore, the use of a structured, top-down approach that starts with a broad matrix, filters by governance, selects based on workload, and concludes by codifying the decision in a boundary contract.

The following table lists the decision matrix for selecting the appropriate architectural pattern:

Use case	Key driver(s)	Recommended pattern	Best for
BI and analytics	Data quality, usability	Cloud DWH + Medallion	Creating reliable, high-quality data products for BI
Regulatory reporting + BI	Audit trail + simple analytics	Medallion + data vault (hybrid)	Regulated industries need both an auditable history and simple analytics
Domain self-service data platform	Domain ownership, autonomy	Data mesh + medallion/ lakehouse	Empowering domains while keeping shared platform standards
Real-time dashboards and alerts	Low latency, high velocity	Streaming lakehouse	Processing continuous data streams for immediate insights
Large-scale periodic reporting	High volume, cost-efficiency	Batch on lakehouse or cloud DWH	Processing large datasets where latency is not critical

***Table 2.2**: Decision matrix*

Concrete example

A global financial institution needs to build a new platform for regulatory reporting that must also support ad-hoc analytics for fraud detection. The platform must ingest millions of trade transactions daily and retain a perfect, auditable history for seven years. The team begins its selection process using the decision matrix at *Table 2.1*, which points them toward medallion and data vault patterns due to the need for both curated analytics and historical tracking.

Following the governance flow (*Figure 2.3*), the strict auditability requirement makes a data vault component essential for tracking the history of every transaction. However, the workload flow (*Figure 2.4*) shows that business analysts require simple, denormalized tables for their BI tools, a key strength of the medallion architecture's gold layer, as illustrated in *Figure 2.1*. Based on the detailed breakdown in *Table 2.2*, the team chooses a hybrid approach. They use a data vault model, whose core components are shown in *Figure 2.2*, within the silver layer for raw data integration and historical tracking, while creating curated gold tables for analytics. The team can deliver new analytics features quickly and still provide regulators with a complete, auditable history, with a link to the version-controlled code that implements the auditable ingestion pattern (*Listing 2.1*) serving as evidence.

Therefore, the institution adopts a hybrid design: a data vault model in the silver layer for long-term, cross-domain history, combined with medallion-style gold tables on a cloud DWH or lakehouse for simple analytics, satisfying both strict regulatory reporting and everyday BI needs.

The following table lists the comparison of medallion-only vs. medallion with data vault:

Criteria	Medallion-only	Medallion with data vault
Auditability needs	Moderate	High
Development simplicity	High	Moderate
Team expertise	Lower	Higher
When to combine medallion and data vault	Short-lived analytics, minimal upserts, limited governance.	Long-term auditability, frequent upserts, strict regulations, and cross-system lineage

***Table 2.3**: Medallion-only vs. medallion with data vault*

Cautions and limits

While a structured selection process is vital, teams must be aware of common pitfalls that can undermine the outcome. It is crucial to avoid the silver bullet fallacy, where one architecture is incorrectly assumed to solve all problems; every pattern has inherent limitations. Similarly, teams must guard against analysis paralysis, where the evaluation process becomes so prolonged that no decision is made. Finally, architecture by committee often produces a diluted and ineffective compromise that fails to meet any objective well.

The architectural decision itself should be treated as a living contract rather than a one-off slide. A lightweight, version-controlled boundary document, for example, a YAML file reviewed via pull requests, helps the project adapt to new information while keeping core, irreversible limits protected.

Therefore, avoid common anti-patterns like seeking a single perfect solution, and treat the architectural decision as a living contract managed under strict version control.

Glossary

The following are the terms that have been discussed in this section:

- **Cloud DWH (cloud data warehouse)**: A fully managed, cloud-native analytical database (for example, Snowflake, BigQuery, or Redshift) that separates storage and compute, scales elastically, and is primarily used for BI and reporting workloads.
- **Boundary contract**: A small, machine-readable document (often YAML) that defines scope, budget, and non-negotiable constraints for a data project.
- **Data vault**: A data modeling methodology designed for long-term historical storage of data from disparate systems. It is optimized for auditability, scalability, and flexibility by separating structural information (Hubs and Links) from descriptive attributes (Satellites).
- **Medallion architecture**: A data architecture pattern that organizes data into three distinct layers: bronze (raw), silver (cleansed and conformed), and gold (aggregated for business analytics). This layered approach progressively improves data quality and usability.
- **Policy-as-code**: The practice of managing and provisioning policies (for security, operations, or architecture) using machine-readable definition files, rather than manual configuration or static documents.

Expert notes

Once an architectural decision is made, its enforcement should be automated to the greatest extent possible. A boundary contract moves beyond being mere documentation when it is integrated into automated workflows. For example, a data quality rule defined in the contract can be wired into your platform's native schema enforcement and data quality tests, so that any data violating the agreed-upon structure or freshness targets causes the pipeline to fail before it reaches curated layers.

Similarly, a requirement for auditable, atomic updates can be enforced by mandating ACID-compliant upsert patterns (such as MERGE-based loads) and by capturing the resulting transaction metadata as part of your audit evidence. Automating these architectural constraints allows developers to build freely within the established boundaries without constant manual oversight. The ultimate assurance artifact is a CI pipeline that automatically runs validation checks and fails any build where a code change violates a rule defined in the boundary contract.

Therefore, once the architecture is chosen, translate its key rules into automated checks, for example, by wiring a boundary contract into CI so that schema, retention, and auditability requirements are enforced by pipelines rather than manual review.

Related topics

This section focuses on architecture selection and decision codification. For related topics addressed elsewhere in this book:

- Medallion architecture implementation, see *Chapter 4, Medallion Architecture in Multi-cloud*
- Data vault implementation, see the *Data vault modeling* section in this chapter
- Cloud infrastructure provisioning, see *Chapter 9, Cloud Migration and Coexistence Strategies*
- Data pipeline tools, see *Chapter 5, Building Scalable Data Pipelines*

Lab: Boundary and RB-30

Now that you have compared foundational architectural patterns, this lab puts theory into practice in a small, controlled world. You will work as a data engineer for Fusion-Mart, a fictional omnichannel retailer, and simulate how a single change flows through a lakehouse-style medallion pipeline. Instead of building a full production system, you will use a tiny dataset and a minimal script to evaluate whether a proposed change respects the boundary gate: Is the scope clear? Is the change unit small enough? Also, can you roll back within the RB-30 window?

Lab environment: All labs run in GitHub Codespaces with a dev container. This lab produces `labs/ch02/artifacts/result.json` containing boundary check status, change unit validation, RB-30 rollback assessment, and accept/reject decisions, ensuring consistent governance evidence regardless of technology.

Note: **The lab uses accept/reject for simplicity. In the book text, we use Go/Retry/Kill to describe the same decision workflow.**

Conclusion

This chapter equipped you to confidently reason about the boundary gate: how to choose an architectural pattern that fits your governance needs and how to express non-negotiable limits in plain, operational terms. You explored three foundational paradigms, data mesh, lakehouse, and medallion, and learned how their strengths and trade-offs map to real-world constraints such as retention, auditability, and workload characteristics rather than to vendor product names. Through the lab, you also saw how even a tiny lakehouse-style pipeline can embody boundary thinking and prepare you for RB-30-capable change management in later chapters.

With this architectural foundation in place, your next step is to design how data actually flows into and across this structure from source systems, through refinement layers, to consumption. The next chapter will guide you through developing effective integration strategies and end-to-end data flows, ensuring your initiatives deliver tangible business value and a strong return on investment.

With its medallion and data vault architecture now in place, Fusion-Mart has established the structural foundation for governed data flows.

Join our Discord space

Join our Discord workspace for latest updates, offers, tech happenings around the world, new releases, and sessions with the authors:

https://discord.bpbonline.com

CHAPTER 3
Data Integration Strategy, Business Impact, and ROI

Introduction

Consider this scenario: A retail company's pricing algorithm ingests corrupted product data, resulting in $2M in lost revenue over a single weekend. The root cause? No automated quality gate caught the malformed records before they reached production.

Ensuring the reliability and trustworthiness of data is paramount in any robust data integration strategy. A critical challenge lies in consistently defining and automating quality contracts, specifically **service level indicators** (**SLIs**) and **service level objectives** (**SLOs**) that serve as essential inputs to the deployment quality gate, as detailed in *Chapter 8, DevOps and CI/CD for Data Engineering*. This chapter explores practical methodologies for establishing these quality contracts, transforming abstract technical metrics into tangible business value. By the end, you will be equipped to connect technical SLIs and SLOs directly to measurable business outcomes and **return on investment** (**ROI**). Furthermore, you will learn to implement automated pre-flight checks for data quality within a **continuous integration** (**CI**) pipeline and effectively log exceptions that necessitate human review before data can proceed through critical governance gates.

Structure

This chapter covers the following topics:

- Identifying and tracking ROI for data initiatives
- Overcoming common integration obstacles
- Business outcomes
- Real-world case studies
- Lab: Integration SLO/SLI check

Objectives

By the end of this chapter, you will be able to translate technical data quality metrics into business language using the KPI conversion recipe to connect SLIs and SLOs to ROI and calculate the financial impact of data quality improvements by measuring revenue lift, cost reduction, and compliance risk mitigation. You will design effective data integration strategies that overcome schema evolution, late-arriving data, and cross-system dependencies, while you implement automated quality checks in integration pipelines that validate data before it reaches downstream systems.

Identifying and tracking ROI for data initiatives

Quantifying the business value of data engineering is essential for justifying investments and aligning technical work with strategic goals. This requires a systematic framework, introduced in this book as the KPI-to-SLO conversion recipe, that connects specific engineering initiatives, like pipeline optimization or data integration, to measurable business outcomes. By establishing this clear causal chain, teams can move beyond discussing technical achievements and instead demonstrate their direct contribution to revenue growth, cost reduction, and operational efficiency. This approach enables data teams to deliver innovative solutions while providing stakeholders with transparent, auditable proof of impact (verifiable records such as logs, timestamps, and approval signatures that can be independently reviewed), often tracked via a central business value dashboard. The result is a data-driven narrative that translates technical effort into a compelling ROI.

Key considerations

To effectively measure the business value of data engineering, several key considerations must be addressed from the outset:

- **Establish clear baselines**: Before initiating any project, you must quantify the current state. This involves measuring key business metrics like customer churn, **average order value** (**AOV**), or the time required for manual data reconciliation. Without these initial benchmarks, it is impossible to prove that your technical improvements caused a positive change. This baseline data provides the foundation for demonstrating progress and calculating ROI even as new data sources are integrated.

- **Attribute value to specific capabilities**: A data platform is composed of many features, and it is crucial to isolate the value contributed by each. For example, a new **master data management** (**MDM**) system might directly reduce marketing waste by eliminating duplicate customer records, while a real-time ingestion feature enables time-sensitive offers that increase sales. Using A/B testing and control groups helps attribute uplift to specific technical actions, ensuring that value is tracked through a versioned data dictionary that clarifies metric definitions over time.

- **Recognize phased and evolving ROI**: The return on a data platform investment is not static; it matures over time. Initial efforts may yield operational efficiencies, such as reduced manual data cleaning or faster report generation. As the platform stabilizes and data quality improves, the focus can shift to more direct business impacts like personalized marketing. In advanced stages, the platform can unlock strategic value by enabling predictive analytics and identifying new market opportunities.

Therefore, to prove value, you must first establish baselines, then attribute specific business outcomes to distinct technical capabilities, and finally, track how ROI evolves from operational efficiency to strategic impact over time, all documented through an auditable evidence record.

Process or steps

Translating data engineering work into quantifiable business value involves a structured, repeatable process:

1. **Identify business objective and define metrics**: Begin by collaborating with stakeholders to identify a critical business problem that data can help solve. This could be reducing customer churn, increasing operational efficiency, or improving marketing effectiveness. Once the objective is clear, define the specific **key performance indicators (KPIs)** that will measure success, such as a 10% reduction in churn rate or a 15% increase in marketing campaign conversion.

2. **Map technical actions to business metrics**: Connect the desired business outcome to the specific data engineering tasks required to achieve it. This creates a clear value chain, as shown in *Figure 3.1*, linking a technical action like building a unified **customer data platform (CDP)** to an intermediate outcome like higher data quality, which in turn drives the target business metric. This mapping ensures that technical work remains aligned with strategic goals and can proceed without disrupting existing analytics workflows.

3. **Establish baselines, measure, and report**: Before implementation, measure the current performance of the target KPIs to establish a baseline. After deploying the data engineering solution, continuously monitor these metrics and compare them against the baseline to quantify the uplift. This data provides the core evidence for the project's ROI, which can be compiled and distributed through an automated report generator that pulls directly from monitoring systems.

Therefore, the process involves defining a business objective, mapping the required technical work directly to that objective's metrics, and then measuring the improvement against a pre-established baseline.

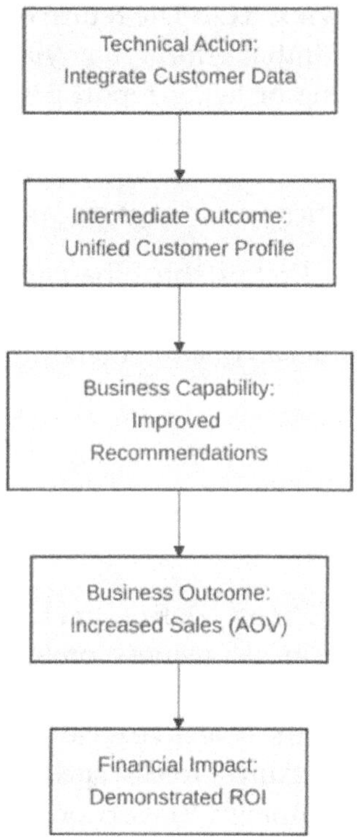

***Figure 3.1**: Simplified value chain*

Concrete example

A retail company, *Fusion-Mart,* sought to increase online sales by improving its product recommendation engine. The existing engine suffered from fragmented customer data, leading to generic suggestions and low conversion rates. Following the defined process, the data engineering team first identified the business objective: increase the AOV by 5%. They mapped this goal to a specific technical action: building a unified customer profile by integrating data from e-commerce, loyalty, and POS systems into a central CDP. Before starting, they recorded the baseline AOV at $50. After launching the new CDP and powering the recommendation engine with unified profiles, they ran an A/B test. The new engine, using high-quality data, delivered more relevant recommendations, increasing the AOV for the test group to $58, a 16% lift. This allowed them to deploy new personalization models with confidence, and still have a clear rollback path managed by a feature flag system. This quantifiable success, translating to millions in additional revenue, directly demonstrated the business value of the data integration project.

Therefore, by integrating fragmented data to create a unified customer profile, Fusion-Mart directly increased its average order value by 16%, providing a clear, measurable return on its data engineering investment.

Cautions and limits

While quantifying the value of data engineering is powerful, it is important to acknowledge its inherent challenges and limitations. Perfect attribution is often impossible, as business outcomes can be influenced by multiple factors simultaneously, such as marketing campaigns or market shifts. Furthermore, there is a risk of focusing on vanity metrics that are easy to measure but do not reflect true business value. The ROI for foundational platform investments also may not be immediate, as initial work focuses on stability and data quality, which are enabling capabilities rather than direct revenue drivers. Teams can mitigate these challenges by using control groups to better isolate impact, focusing on metrics that are causally linked to outcomes, and communicating the phased nature of data platform ROI to stakeholders. It is possible to acknowledge these limitations in a formal decision log and still provide clear, directional evidence of value. Building resilient data integration workflows, also helps manage technical risks that could otherwise obscure business impact.

Therefore, perfect attribution is difficult and ROI can be delayed, so teams must mitigate these challenges by using control groups, focusing on meaningful metrics, and setting realistic stakeholder expectations.

Glossary

Here are a few key terms that are discussed in this section:

- **Average order value (AOV)**: A sales metric that measures the average total of every order placed with a merchant over a defined period.
- **Customer data platform (CDP)**: A system that collects and unifies customer data from multiple sources to build a single, coherent view of each customer.
- **Key performance indicator (KPI)**: A quantifiable measure used to evaluate the success of an organization, employee, or project in meeting objectives for performance.
- **Master data management (MDM)**: A technology-enabled discipline in which business and IT work together to ensure the uniformity, accuracy, and consistency of an organization's critical data assets.
- **Return on investment (ROI)**: A performance measure used to evaluate the efficiency or profitability of an investment. It is calculated by dividing the net profit of an investment by its cost.

Expert notes

For a more sophisticated analysis, teams can move beyond simple before-and-after comparisons to more advanced attribution models. For instance, linking technical SLOs, such as data freshness or pipeline uptime, directly to business KPIs provides a continuous, real-time view of impact. A framework like the one detailed in the KPI conversion recipe (*Table 3.1*) can formalize this connection. This allows teams to calculate the financial cost of a missed SLO, such as the revenue lost when a data pipeline delay prevents a time-sensitive marketing campaign from running. This advanced measurement allows teams to refine complex data models while ensuring core data pipelines remain stable and reliable. Implementing a mandatory code review checklist that includes checks for metric instrumentation ensures that this level of observability is built into the development process from the start.

Therefore, advanced teams can directly connect technical SLOs to financial KPIs, enabling them to calculate the precise business cost of data quality issues or pipeline failures.

KPI	SLI	SLO	Business impact
Data freshness	*MAX(load_timestamp) - MAX(event_timestamp)*	< 1 hour	Enables near-real-time inventory management and timely customer personalization, preventing lost sales opportunities
Data quality	*(Valid Records / Total Records) * 100*	> 99.9%	Guarantees trustworthy analytics, avoiding flawed business decisions and loss of customer trust
Pipeline reliability	*(Successful Runs / Scheduled Runs) * 100* (over 30 days)	> 99.5%	Prevents stale data in critical dashboards and services, maintaining trust in the data platform

***Table 3.1**: KPI conversion recipe*

Related topics

This section focuses on quantifying the business value of data engineering projects. For related topics addressed elsewhere in this book:

- ETL/ELT pipeline implementation, see *Chapter 5, Building Scalable Data Pipelines*
- Data governance and organizational alignment, see *Chapter 6, Data Governance and Compliance*

Overcoming common integration obstacles

Modern data integration requires a systematic approach to ensure data quality and manage complex dependencies. The core challenge is transforming abstract quality requirements into concrete, automated enforcement mechanisms. This is achieved by establishing clear quality contracts, implementing automated pre-flight validation checks within a CI pipeline, and using sophisticated orchestration patterns defined as **infrastructure as code** (**IaC**). This approach enables rapid data processing while ensuring that only high-quality, validated data proceeds to critical systems. A version-controlled data quality ruleset, stored in a Git repository, serves as a concrete assurance artifact for these automated checks, making the entire process auditable and repeatable.

Key considerations

Successful data quality automation hinges on several core principles that connect technical execution to business objectives. These considerations form the foundation for building resilient and trustworthy data pipelines:

- **Translate quality into business value**: Data quality metrics must be directly linked to business outcomes. By defining SLIs, which are technical measures of quality (e.g., percentage of records with a valid `email_address`), and SLOs, which are targets for those SLIs (e.g., 99.9%), teams can quantify the impact of data integrity. This translation, detailed in the KPI conversion recipe shown in *Table 3.1*, visualizes the value chain from a technical action to a business result, as illustrated in *Figure 3.1*, making the ROI clear.
- **Automate validation early**: Integrating automated quality checks early in the data lifecycle, a shift-left approach, meaning data quality validation is moved upstream closer to data creation, rather than fixing issues downstream, is critical for efficiency. Pre-flight checks executed within a CI pipeline catch issues like schema drift or invalid data types before they contaminate downstream systems. This proactive validation significantly reduces the cost and complexity of remediation, preventing data errors from propagating through the analytics stack.
- **Orchestrate with code**: Orchestrate with code: Defining data workflows using IaC provides repeatability, versioning, and auditability. Workflow orchestration tools (such as Apache Airflow, AWS Step Functions, or Azure Data Factory) allow teams to accelerate development with declarative workflows while maintaining consistent quality standards across all environments.

Therefore, effective data integration relies on linking measurable SLIs or SLOs to business ROI, embedding automated pre-flight checks early in the pipeline, and managing workflows as version-controlled IaC.

Process or steps

Implementing a robust data quality framework involves a structured process that moves from definition to automated enforcement and logging:

1. **Define quality contracts**: The first step is to translate business requirements into precise, measurable technical contracts. This involves identifying critical data elements and defining the SLIs and SLOs that govern their quality. For example, for a `customer` dataset, an SLO might state that 99.95% of records must have a non-null `customer_id` that exists in the master `accounts` table.

2. **Implement automated checks**: Convert the defined quality contracts into executable code within your data pipeline. These automated pre-promotion checks should cover schema validation, data type enforcement, uniqueness, and referential integrity. These checks act as automated gates, ensuring data conforms to its contract before it can be promoted to a production environment.
3. **Log and route outcomes**: The final step is to act on the validation results and record them for auditability. The pipeline's orchestration logic should conditionally route data: valid data proceeds to the next stage, while invalid data is sent to an error-handling workflow for quarantine or remediation. Every decision must be recorded using a structured format to automate the data flow without sacrificing auditability. An immutable log stream in Amazon CloudWatch that records every validation decision serves as a durable assurance artifact.

Therefore, the process for ensuring data quality follows a clear cycle of defining contracts as SLIs or SLOs, implementing them as automated pre-flight checks, and logging all outcomes to guide the data flow.

Concrete example

An e-commerce platform can use a workflow orchestration tool to automate quality checks for new customer sign-up data. When a new customer record is written to cloud object storage, an event triggers a workflow that orchestrates the validation process. The workflow begins by invoking a serverless function that acts as a pre-flight check, validating the incoming data against a predefined schema: it verifies that the email field is correctly formatted, the `postal_code` matches a valid pattern, and the age is within a reasonable range (e.g., 18 to 120).

A conditional branch within the workflow then evaluates the function's output. If the data passes all checks, it is forwarded to a processing job for transformation and loading into a data warehouse for analysis. However, if any validation rule fails, the workflow routes the record to a different path. The invalid record is moved to a separate quarantine bucket, an entry detailing the specific validation failure is written to a centralized logging service, and a notification is sent to a data stewardship team for manual review. This allows the platform to process thousands of records per minute while automatically quarantining every invalid record. A feature flag that can disable the data loading step without redeploying the entire pipeline provides a concrete assurance artifact for operational control.

Therefore, a workflow orchestration tool can effectively serve as a data quality gatekeeper by orchestrating a workflow that validates incoming records, processes valid data, and automatically quarantines and logs invalid data for review.

Cautions and limits

While automated quality contracts are powerful, their implementation requires careful consideration of potential challenges to be successful:

- **Overly aggressive SLOs**: Setting SLOs too close to 100% without understanding natural data variance can create excessive noise. This can lead to valid data being incorrectly quarantined, blocking critical business processes. To mitigate this, start with more lenient SLOs based on an initial data profile, and tighten them incrementally as data quality improves and processes stabilize.
- **Performance impact**: Each validation check adds processing overhead and latency to a data pipeline. Complex checks, especially those requiring lookups against large datasets for referential integrity, can slow down data availability. Mitigation involves optimizing validation code, running independent checks in parallel, and using fit-for-purpose tools to accelerate performance while maintaining rigor.
- **Schema and contract drift**: Business requirements and data sources evolve, causing schemas and quality rules to become outdated. Without a management process, these contracts can become ineffective. To manage this, treat quality contract definitions as code, version them in a repository alongside the pipeline code, and use a schema registry that enforces backward compatibility rules on all changes to evolve data contracts quickly without disrupting downstream consumers.

Therefore, to be effective, data quality contracts must be practical, performance-aware, and actively managed as versioned code to adapt to evolving data landscapes.

Glossary

Here are a few key terms that are discussed in this section:

- **IaC**: The practice of managing and provisioning infrastructure through machine-readable definition files, rather than through physical hardware configuration or interactive configuration tools.
- **State machine**: A workflow orchestration pattern that coordinates multiple processing steps through defined states and transitions. Implementations include Apache Airflow, AWS Step Functions, Azure Data Factory, and Prefect.
- **Schema drift**: An issue that occurs when the structure of incoming data changes over time, causing it to no longer align with the expected schema defined in the data pipeline or target database.

Expert notes

For teams seeking to advance their data quality practices, several advanced concepts can enhance resilience and adaptability:

- **Design for idempotency**: Ensure that data processing tasks are idempotent, meaning that running a task multiple times with the same input produces the same result. This is crucial for building reliable error-handling and retry mechanisms, as a failed task can be safely re-executed without causing data duplication or corruption.

- **Adopt error budgets**: An error budget, derived from your SLO (e.g., a 99.9% SLO yields a 0.1% error budget), provides a data-driven framework for balancing innovation and reliability. Teams are free to deploy new features and make changes as long as the resulting errors do not exhaust the budget. This allows teams to innovate with advanced data models and still maintain a stable and predictable data platform. A documented error budget policy that is reviewed quarterly by the platform and business teams is a key assurance artifact.
- **Explore dynamic anomaly detection**: Instead of relying solely on static, predefined rules, consider using machine learning models to detect anomalies in data streams. These models can learn the normal patterns and distributions of your data and flag significant deviations as potential quality issues, enabling the detection of unknown or unexpected problems that static rules might miss.

Therefore, mature data quality systems move beyond static rules by incorporating idempotent design for safe retries, using error budgets to balance innovation with reliability, and leveraging anomaly detection for dynamic monitoring.

Related topics

This section focuses on defining and automating data quality contracts. For related topics addressed elsewhere in this book:

- Data lineage and cataloging, see *Chapter 6, Data Governance and Compliance*
- CI/CD pipeline configuration, see *Chapter 8, DevOps and CI/CD for Data Engineering*
- Master data management, see *Chapter 6, Data Governance and Compliance*

Business outcomes

The success of a data project is measured by its impact on business outcomes, not by its technical sophistication alone. While system performance metrics are important, their true value emerges only when linked to measurable improvements in revenue, operational efficiency, or strategic goals. This requires a shift from a purely technical viewpoint to a framework that connects execution directly to business value. This connection is forged through the disciplined application of KPIs, which define business success; SLIs, which measure technical performance; and SLOs, which set the quality targets that technical work must meet to achieve the desired business results. This approach allows teams to innovate rapidly while ensuring data quality standards are consistently met, with all quality checks and approvals tracked in a version-controlled decision log.

Key considerations

To effectively link technical data operations with strategic business goals, several core principles must be established. These considerations form the foundation of a data quality framework that is both robust and aligned with organizational objectives:

- **Focus on business value**: The ultimate measure of any data initiative is its contribution to tangible business outcomes. Technical metrics like pipeline execution time or processing speed are secondary to high-level KPIs, such as customer retention or operational cost reduction. Therefore, all technical efforts must be justified by their direct or indirect support for a specific business KPI.
- **Establish a clear hierarchy**: The relationship between business goals and technical metrics must be explicit and hierarchical. Business-centric KPIs sit at the top, defining strategic success. These are supported by SLOs, which are specific, measurable targets (e.g., 99.9% data completeness) that, if met, ensure the business goal is achievable. SLOs are themselves measured by technical SLIs, such as the actual percentage of complete records.
- **Prioritize data reliability**: The entire framework depends on the quality and trustworthiness of the underlying data. Flawed data leads to flawed insights and poor business decisions, rendering KPIs and SLOs meaningless. Foundational data quality rules, covering completeness, format, range, uniqueness, and referential integrity, are the non-negotiable bedrock upon which reliable metrics are built.

Teams can pursue aggressive development goals while automated quality checks provide a safety net, and a feature flag allows for the immediate rollback of a data pipeline change if it negatively impacts a key SLO.

Therefore, a successful data quality strategy requires translating high-level business KPIs into specific, measurable technical SLOs that are built upon a foundation of fundamental data reliability rules.

Process or steps

Connecting technical work to business outcomes involves a systematic process of defining and monitoring quality contracts:

1. **Define business KPIs**: The process begins with a clear business objective (the cause). Stakeholders must define a high-level, strategic metric that represents success, such as increasing sales or reducing customer churn (the action). The result is a specific, measurable KPI that provides a clear target for all subsequent data work, a concept detailed in the KPI conversion recipe at *Table 3.1*.
2. **Identify technical SLIs**: With a KPI defined (the cause), the next step is to identify the underlying technical conditions required to achieve it. Data teams must determine

which quantifiable aspects of the data or service, such as freshness, accuracy, or completeness, are critical (the action). The result is a set of specific SLIs that can be measured continuously to gauge the health of the data platform.

3. **Establish SLOs and automate checks**: For each critical SLI (the cause), the team must set a precise, achievable target that aligns with business expectations, creating an SLO. This SLO is then translated into an automated pre-flight check within the CI pipeline (the action). The result is a reliable deployment gate that programmatically enforces data quality standards before data is used in production, preventing low-quality data from causing business disruption.

This structured process enables the continuous delivery of data products and still provides stakeholders with verifiable proof of data integrity through an immutable audit trail that logs every automated check and its outcome. This audit trail, which records every decision and its evidence, becomes the foundation for the RB-30 rollback capability introduced in *Chapter 5, Building Scalable Data Pipelines*. Therefore, the process involves translating a strategic business KPI into measurable technical SLIs, then enforcing quality by setting SLO targets as automated pre-flight checks in a CI pipeline.

Concrete example

Consider an e-commerce company whose primary business KPI is to *increase customer lifetime value by 15%*. The strategy to achieve this involves highly personalized email marketing campaigns, which depend entirely on the quality of the customer dataset. The team translates this business need into a technical quality contract. The marketing team can launch new campaigns frequently, even as the automated SLO validation ensures the underlying data meets strict quality criteria. A dedicated quarantine table automatically isolates records that fail the completeness check, preventing them from entering the production marketing dataset.

The team identifies **`customer_profile_completeness`** as a critical SLI, measuring the percentage of customer records that contain a valid email address, a full name, and at least one purchase history entry. After consulting with business stakeholders, they establish a SLO that 99.5% of records in the **`production.customers`** table must meet the completeness criteria.

This SLO is implemented as an automated pre-flight check in the CI pipeline that runs before the weekly marketing audience is generated. The check executes a query against the data warehouse to calculate the **`customer_profile_completeness`** SLI. If the result is 99.5% or higher, the pipeline proceeds. If it falls below 99.5%, the pipeline fails, an alert is triggered, and the flawed data is prevented from being used for the marketing campaign, thus safeguarding the business KPI.

Therefore, an e-commerce company protects its KPI for customer lifetime value by creating an SLO for 99.5% customer profile completeness and enforcing it with an automated CI pipeline check.

Cautions and limits

While implementing a KPI-driven data quality framework is powerful, teams must be aware of potential pitfalls and limitations. These challenges are manageable with foresight and proper planning. Organizations can adopt this quality framework incrementally without halting existing workflows, and still build confidence in their data assets over time. A documented escalation path ensures that any SLO breach that is not automatically resolved triggers an alert to the on-call data engineer.

One common risk is setting misaligned KPIs that do not reflect true business value. If the top-level metric is flawed, the entire system of SLIs and SLOs will optimize for the wrong outcome. To mitigate this, business stakeholders must be deeply involved in defining and validating KPIs. Another challenge is over-engineering SLOs by setting targets that are unnecessarily strict or covering too many trivial metrics, which can slow down development. Teams should start with a few critical SLOs tied to the most important data assets and iterate from there.

Finally, a process for handling exceptions is essential. Not all data that fails an SLO check is irrelevant; it may require manual review and correction. Without a clear workflow for managing these exceptions, bad data can be ignored or, conversely, block critical processes indefinitely. Implementing a state machine for handling retries and quarantining bad data, provides a resilient pattern for managing these failures gracefully.

Therefore, to avoid common pitfalls, ensure KPIs are aligned with true business value, start with a few critical SLOs to avoid over-engineering, and implement a clear exception-handling workflow for data that fails validation.

Glossary

Here are a few key terms that are discussed in this section:

- **SLI**: A specific, quantifiable technical metric of the performance or quality of a service. Examples include data freshness, system uptime, or error rate. SLIs are the raw measurements used to evaluate SLOs.
- **SLO**: A target value or range for an SLI that is agreed upon by stakeholders. An SLO represents the desired level of quality or performance and serves as a formal contract between the data team and the business.
- **CI**: A software development practice where developers frequently merge their code changes into a central repository, after which automated builds and tests are run. In a data context, CI pipelines are used to automate data quality checks.
- **Data quality**: The measure of data's fitness to serve its specific purpose. It is commonly assessed against dimensions such as completeness, format, range, uniqueness, and referential integrity.

Expert notes

For experienced practitioners, the KPI/SLI/SLO framework can be extended with more advanced concepts to further enhance reliability and innovation. A key concept is the Error Budget, which is the mathematical inverse of an SLO (e.g., a 99.9% uptime SLO implies a 0.1% error budget). This budget gives development teams a clear, quantifiable allowance for risk and failure, empowering them to innovate and deploy new features rapidly if they remain within the budget. Advanced teams can implement dynamic SLOs that adjust to business seasonality while maintaining a consistent, auditable record of data quality, managed through a version-controlled SLO definition file where changes are approved via pull requests.

This framework visualizes the direct value chain from a technical action, like data integration, to a business outcome, such as increased sales, as shown in *Figure 3.1*.

For complex data workflows involving multiple stages of validation, enrichment, and transformation, a formal state machine becomes invaluable. By explicitly defining states for processes like validation pending, quarantined, and retry, teams can build highly resilient and observable pipelines using workflow orchestration tools of their choice.

Therefore, advanced teams can use error budgets to balance innovation with reliability and employ state machines to build resilient, observable data pipelines for complex, multi-stage workflows.

Related topics

This section focuses on the KPI/SLI/SLO framework for data quality contracts. For related topics addressed elsewhere in this book:

- CI/CD tool selection and configuration, see *Chapter 8, DevOps and CI/CD for Data Engineering*
- Data ownership and access control, see *Chapter 6, Data Governance and Compliance*
- Data lineage tracking, see *Chapter 6, Data Governance and Compliance*

Real-world case studies

Demonstrating the tangible impact of robust data quality strategies is paramount to proving their value beyond theoretical frameworks. This article provides concrete evidence of how technical implementations translate into measurable business outcomes across diverse industries. By examining case studies from retail, financial services, and healthcare, we uncover universal patterns of success that connect automated data quality enforcement to quantifiable ROI. These examples show that a systematic approach to data governance is not an operational burden but a direct driver of revenue, efficiency, compliance, and safety, enabling rapid innovation without sacrificing stability. A central, auditable evidence log provides the necessary proof of these controls.

Key considerations

Before implementing an automated data quality framework, it is essential to understand the core principles that drive its success. These considerations form the foundation for a strategy that delivers reliable data and measurable business value.

- **Proactive quality enforcement**: The most effective data quality strategies operate proactively, not reactively. By codifying quality rules into automated pre-flight checks, organizations can prevent flawed data from ever entering their systems. This shift-left approach moves quality validation to the earliest point in the data lifecycle, which minimizes the high cost of downstream remediation and allows development to proceed quickly while still protecting system integrity.
- **Connecting technical metrics to business value**: Data quality for its own sake is an academic exercise; its value is realized only when tied to business outcomes. It is critical to translate business KPIs into technical SLOs and SLIs. This process, outlined in the KPI conversion recipe at *Table 3.1*, creates a clear value chain from a technical action to a business result, as visualized in *Figure 3.1*, justifying investment and aligning technical teams with business goals.
- **Centralized and auditable evidence**: A cornerstone of robust governance is the ability to prove compliance and demonstrate control. Consolidating all data quality check results, lineage information, and approval decisions into a single, auditable audit trail is non-negotiable. This centralized repository, detailed in *Chapter 6, Data Governance and Compliance*, enables one-click evidence retrieval and provides a transparent, immutable record of data integrity, even as the data itself is processed and transformed.

Therefore, a successful data quality strategy hinges on proactive enforcement through automated checks, a clear link between technical metrics and business ROI, and a centralized, auditable evidence trail for governance.

Process or steps

Implementing a systematic data quality and governance framework involves three key steps:

1. **Define and translate objectives**: The process begins with a business problem, such as a high fraud rate or low customer conversion. This business-level KPI must be translated into a measurable technical SLO, such as *fraud detection accuracy > 99.5%*, which is in turn measured by specific SLIs like *transaction data completeness > 99.9%*. This crucial translation, guided by resources like *Table 3.1*, ensures that technical efforts directly serve and measurably impact business goals.
2. **Codify and automate quality contracts**: With clear SLIs defined, the next step is to codify these requirements into automated data quality contracts. These contracts specify rules for data completeness, format, range, uniqueness, and referential

integrity. They are implemented as automated pre-flight checks within a CI pipeline, ensuring that every new data source or transformation is validated before integration. This automation prevents defective data from propagating downstream, protecting the reliability of analytics and operational systems.

3. **Implement a governed decision workflow**: The results from the automated checks must feed into a structured decision process to manage risk effectively. This workflow, often implemented as a state machine, determines the fate of each data batch: Go if all checks pass, Retry for minor, correctable issues, or Kill for critical failures. Each decision is logged with supporting evidence, creating an audit trail that can be referenced via a permanent URL, providing clear accountability for data promotion.

Therefore, the process involves translating business needs into technical SLIs, automating quality contract enforcement in a CI pipeline, and using a governed workflow to approve, remediate, or reject data based on the results.

Concrete example

Fusion-Mart, an e-commerce retailer, provides a practical illustration of this process. The company struggled with fragmented customer data, which led to ineffective marketing and poor personalization (the cause). To address this, they first defined a business goal of increasing personalized recommendation effectiveness. This was translated into a technical SLO: achieve a **personalized recommendation conversion rate** (**PRCR**) increase of 15%, supported by an SLI of *customer profile completeness (CPC) > 98%*.

Next, Fusion-Mart codified data quality contracts for its customer data platform. These contracts, enforced via automated pre-flight checks in their CI pipeline, validated incoming data for mandatory fields like `email` and `primary_address`, correct email format using regular expressions, and the uniqueness of the `customer_ID`. This automated enforcement ensured that only high-quality, complete profiles entered their core system, allowing them to build a reliable single view of the customer without manual intervention. A version-controlled YAML file defining these contracts served as the single source of truth for quality rules.

Finally, they implemented a Go/Retry/Kill decision workflow. A data batch with 99% profile completeness would Go to production. A batch with minor, fixable formatting issues might Retry after an automated cleanup script runs. However, a batch with a significant number of missing email addresses would trigger a Kill decision, preventing the low-quality data from degrading their personalization engine. This structured process directly resulted in a 17% increase in conversion rates and a $12 million annual revenue uplift.

Therefore, by translating a business need for better personalization into a technical SLI for data completeness, Fusion-Mart used automated contracts and a decision workflow to improve data quality and achieve significant revenue growth.

Cautions and limits

While powerful, implementing an automated data quality framework requires careful consideration of its limitations and potential challenges. One primary concern is the initial implementation overhead; defining SLIs, building CI checks, and configuring workflows demands a significant upfront investment of time and resources. To manage this, organizations should start with a single, high-value data domain to prove the concept and demonstrate ROI before a broader rollout. This allows teams to deliver value quickly while building momentum for the program.

Another challenge is the risk of creating overly rigid quality contracts that block useful, albeit imperfect, data. A contract that rejects an entire dataset due to a few missing optional fields can halt business processes unnecessarily. The Retry path in the decision workflow is a key mitigation, allowing for manual review or automated remediation of non-critical issues. An exception log provides a crucial assurance artifact by documenting every override and the justification for it.

Finally, this framework represents a cultural shift toward shared data ownership, which can face resistance. Technical teams must work closely with business stakeholders to define meaningful quality metrics and demonstrate the connection between data integrity and business outcomes, as shown in *Figure 3.1*. Without this collaboration, the framework may be seen as a technical barrier rather than a business enabler.

Therefore, organizations must manage the initial implementation cost, design flexible contracts with clear exception paths, and foster a culture of shared data responsibility to overcome the framework's inherent challenges.

Glossary

Here are a few key terms that are discussed in this section:

- **CI**: An automated software development practice where developers frequently merge their code changes into a central repository, after which automated builds and tests are run.
- **KPI**: A quantifiable measure of performance overtime for a specific business objective.
- **Payment Card Industry Data Security Standard (PCI DSS)**: A set of security standards designed to ensure that all companies that accept, process, store, or transmit credit card information maintain a secure environment.
- **Protected Health Information (PHI)**: Any information in a medical record that can be used to identify an individual, and that was created, used, or disclosed in the course of providing a health care service.
- **Personally identifiable information (PII)**: Any data that could potentially identify a specific individual. Examples include name, address, and email address.

- **SLI**: A specific, quantitative measure of some aspect of the level of service being provided. For example, the data completeness percentage.
- **SLO**: A target value or range of values for a service level that is measured by an SLI. For example, achieving 99.9% data completeness over a month.

Expert notes

For experienced practitioners, the value of this framework can be extended with more advanced techniques. The Go/Retry/Kill workflow is fundamentally a state machine, which can be robustly implemented using workflow orchestration tools. Mapping the workflow steps to specific constructs in your chosen tool provides a scalable and observable architecture for data processing that moves projects forward while ensuring data integrity.

Furthermore, data quality contracts should not be static. As business requirements evolve, these contracts must be versioned and updated through a governed process. Storing data quality rules in a version-controlled repository (e.g., Git) alongside application code treats data quality as a first-class citizen of the development lifecycle. This ensures that changes to quality rules are deliberate, auditable, and aligned with software updates. The commit history for these rule files becomes a critical assurance artifact for tracking the evolution of data governance policies.

Finally, integrating comprehensive data lineage into the audit trail provides deeper insights for root cause analysis. When a data quality issue is detected, lineage information allows teams to trace the problematic data back to its source system and pinpoint the exact transformation step where the error was introduced. This capability is invaluable for rapid troubleshooting and preventing the recurrence of similar issues.

Therefore, advanced implementations should leverage state machines for workflow automation, manage quality contracts under version control, and integrate data lineage for enhanced traceability and root cause analysis.

Related topics

This section focuses on the strategic implementation and business value of automated data quality frameworks. For related topics addressed elsewhere in this book:

- Data modeling principles, see *Chapter 2, Understanding Data Mesh, Lakehouse, and Medallion*
- Master data management strategies, see *Chapter 6, Data Governance and Compliance*
- Compliance and audit requirements, see *Chapter 6, Data Governance and Compliance*

Lab: Integration SLO/SLI check

In this hands-on lab, your objective is to evaluate an integration plan against pre-defined SLI/SLO and ROI targets, using a guard pattern. You will specifically focus on verifying that the proposed plan meets the data quality contracts (SLIs/SLOs) detailed in our KPI recipe table (*Table 3.1*). By the end of this lab, you will have validated a foundational pipeline plan that automates quality enforcement and confirms technical SLIs align with tangible business outcomes.

Lab environment: All labs run in GitHub Codespaces with a dev container. The lab produces `labs/ch03/artifacts/result.json` containing validation status, SLI/SLO check results, ROI metrics, and accept/reject decisions, ensuring consistent governance evidence regardless of technology.

Note: **The lab uses accept/reject for simplicity. In the book text, we use Go/Retry/Kill to describe the same decision workflow.**

Conclusion

This chapter demonstrated how defining and automating quality contracts, through SLIs and SLOs, directly translates technical rigor into tangible business value and ROI. You have learned to implement automated pre-flight data quality checks within your CI pipeline, ensuring robust data integrity, and established mechanisms to log exceptions for human review, safeguarding the deployment gate from critical errors.

By mastering these principles, you are now equipped to build resilient data integration strategies that not only meet technical specifications but also drive significant organizational impact. The ability to proactively manage data quality and its associated risks is paramount, empowering you to foster trust in your data assets and accelerate informed decision-making across your enterprise.

In the next chapter, we will shift our focus from data integration strategy to the architectural structure that fundamentally guarantees the quality and consistency we have defined. You will learn to design a concrete medallion architecture blueprint, specifically tailored to operate within boundary gate constraints, and map its bronze, silver, and gold layers to specific cloud services, thereby establishing the robust foundation for high-quality data and secure multi-cloud operations.

Fusion-Mart's achievement, a 16% AOV lift from $50 to $58, demonstrates how quality contracts translate directly into revenue growth.

CHAPTER 4
Medallion Architecture in Multi-cloud

Introduction

How do you maintain governance when your data spans multiple cloud providers? Organizations often struggle to implement medallion architecture consistently across multiple cloud providers while maintaining security and governance. Without a clear blueprint, teams face fragmented data pipelines, inconsistent access controls, and audit gaps that complicate compliance.

Navigating the complexities of multi-cloud environments while establishing robust data governance presents a significant challenge for modern data platforms. Specifically, designing a concrete medallion architecture blueprint that operates effectively within the constraints of the boundary gate, as introduced in *Chapter 2, Understanding Data Mesh, Lakehouse, and Medallion*, requires careful consideration of architectural patterns and security implications. This chapter provides a practical framework to address these challenges, guiding you through the strategic decisions necessary to implement a scalable and secure multi-cloud data solution. By the end, you will be equipped to map medallion layers to specific cloud services for the Fusion-Mart case study, design a secure cross-cloud data access pattern, and document the trade-offs of key multi-cloud implementation choices, ensuring your architecture meets both performance and compliance requirements.

Structure

This chapter covers the following key topics:

- Medallion architecture layers
- Multi-cloud deployment
- IAM and logging strategies in multi-cloud
- Delta Lake and other lakehouse table formats for multi-cloud
- Artificial intelligence pipeline integration
- Lab: Medallion consistency check

Objectives

By the end of this chapter, you will be able to design medallion architecture with Bronze, Silver, and Gold layers that progressively refine data from raw ingestion to business consumption and map medallion layers to cloud services across AWS, Azure, and GCP while maintaining consistent governance regardless of provider. You will implement multi-cloud deployments that leverage platform strengths while avoiding vendor lock-in and maintaining quality standards, applying boundary constraints to multi-cloud architectures, ensuring decisions respect budget limits, security policies, and regulatory requirements. Finally, you will validate data quality as it flows between medallion layers using automated checks that prevent corrupted data from reaching consumers.

Medallion architecture layers

The medallion architecture provides a structured methodology for refining data, acting as a foundational *shield of trust* within a secure data boundary. This pattern organizes data processing into three distinct logical layers, such as bronze, silver, and gold, to progressively enhance data quality, governance, and security from initial ingestion to final consumption. By making all data movement explicit and auditable, this architecture ensures that information is reliable and trustworthy. We will use Fusion-Mart's personalization and fraud detection data pipelines as a running example to illustrate these concepts in a practical, multi-cloud context.

Note: **This chapter focuses on technical governance, the automated controls, policies, and artifacts that enforce data quality and security. Organizational governance aspects such as roles, ownership, and escalation procedures are covered in Chapter 6, Data Governance and Compliance.**

Key considerations

Before implementing the architecture, it is crucial to understand its core principles. These considerations ensure the framework is applied effectively to deliver reliable, secure, and high-quality data. The following points highlight the foundational concepts that drive a successful medallion architecture deployment:

- **Logical vs. physical separation**: The bronze, silver, and gold layers represent logical distinctions in data quality, not necessarily separate physical storage systems. A single cloud service, such as an object store like Amazon **Simple Storage Service (S3)**, can host all three layers, using distinct prefixes or buckets to maintain separation. This approach allows organizations to optimize costs and simplify infrastructure management while still enforcing strict governance and quality gates between the layers, as shown in the service mapping in *Figure 4.1*.
- **Immutable raw data archive**: The bronze layer's primary function is to be an immutable (write-once, never-change) archive of source data. Capturing data in

its original, raw format creates a permanent source of truth. This immutability is critical for auditability; inspectors can verify the original data hash, timestamp, and source metadata at any time. If an audit check fails (e.g., missing lineage or tampered records), the system triggers a Retry workflow: the affected data is quarantined, and downstream processing is paused until the issue is resolved. This allows for the complete reprocessing of downstream silver and gold tables if business logic changes or errors are discovered, ensuring data lineage is always traceable to its origin.

- **Pervasive security and governance**: Security is not an afterthought, but an integral component woven through every layer. Access controls must enforce the **Principle of Least Privilege (PoLP)** and become progressively stricter as data moves from bronze to gold, ensuring that raw data is tightly controlled while curated insights are more broadly but securely accessible. This is achieved by building a secure cross-cloud read path, which combines centralized identity, short-lived credentials, and private networking, as illustrated in *Figure 4.2*, to protect data at every stage.

This logical separation allows teams to innovate on data models in the gold layer without disrupting the foundational data ingestion processes in the bronze layer. A version-controlled schema registry, which requires a formal pull request to approve changes, serves as a concrete assurance artifact to prevent breaking changes in the silver layer. Therefore, the architecture's success hinges on understanding the layers as logical quality gates, preserving raw data immutability, and integrating security from the start.

The following figure shows the mapping of the logical medallion layers to concrete services in a cloud environment (e.g., AWS):

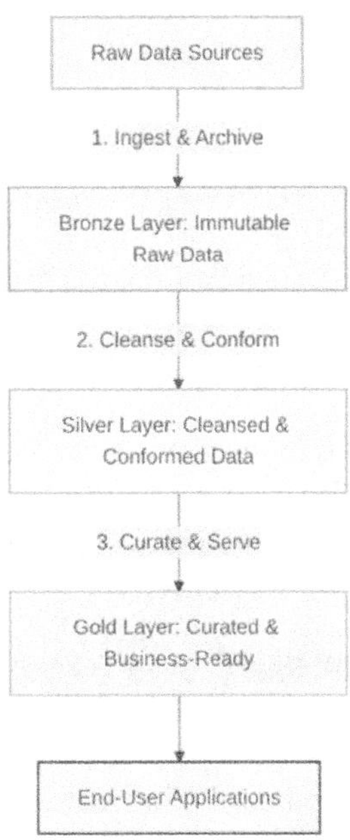

***Figure 4.1**: Mapping the logical medallion layers*

The following figure shows a cross-cloud secure read path (IdP I STS I PE I Table):

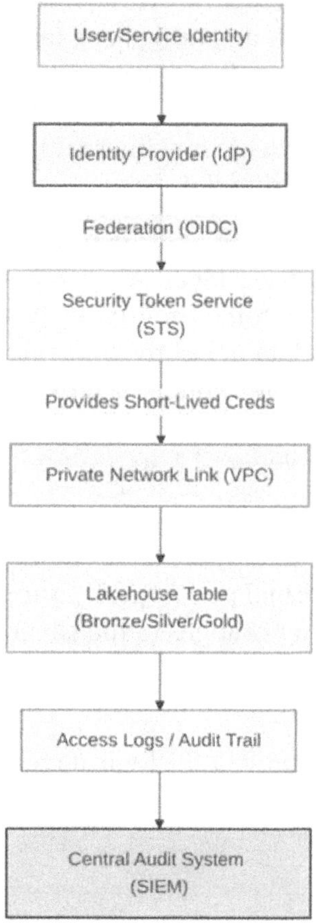

***Figure 4.2**: Cross-cloud secure read path*

Process or steps

Refining raw data into business-ready insights follows a systematic, three-stage process.

1. **Ingest and archive**: Raw data from various source systems is the trigger for the process. The data is ingested into the bronze layer with its original schema and format intact, creating an immutable, auditable historical record for future use. This step prioritizes capture fidelity over immediate usability.

2. **Cleanse and conform**: The availability of raw data in the bronze layer initiates the next action. Processing engines read this data, apply cleansing rules, parse nested structures, enrich it with other enterprise data, and standardize it into a consistent, conformed model in the silver layer. The result is a high-quality, integrated dataset that serves as a reliable foundation for multiple business use cases.

3. **Curate and serve**: Business requirements for specific analytics or applications drive the final step. Data from the silver layer is aggregated, transformed into performance-optimized data marts (e.g., star schemas), and loaded into the gold layer. This produces secure, business-validated data ready for direct consumption by **business intelligence** (**BI**) tools, machine learning models, and end-user applications.

This structured flow ensures data quality improves at each stage, and still provides the flexibility to introduce new data sources by simply adding them to the bronze ingestion step. The data processing framework provides assurance by automatically generating a data lineage graph, which serves as an auditable record of every transformation from source to final product. Therefore, the process systematically transforms raw, untrusted data into validated, business-ready insights through sequential stages of ingestion, cleansing, and curation.

Concrete example

Let us trace a user interaction through Fusion-Mart's personalization pipeline to make this process tangible. The journey begins when a customer clicks on a product on the mobile app, generating a raw event that must be processed to influence future product recommendations.

First, the raw clickstream event, a semi-structured JSON object, is captured and lands in the bronze layer object store. Here, it is stored exactly as it was received, with all its original fields, including a nested **`device_info`** object and a raw user-agent string. This preserves the original event for auditing and potential future reprocessing.

Next, a scheduled batch job reads new JSON files from the bronze layer. This job, which could be a Spark process running in **Google Cloud Platform** (**GCP**), securely accesses the data in **Amazon Web Services** (**AWS**) S3. It parses the JSON, extracts the **`device_type`** and **`browser`** from the user-agent string, joins the data with a **`customer`** dimension table to resolve the user's **`CustomerID`**, and standardizes the **`productID`**. The output is a clean, structured, and enriched table in the silver layer representing user sessions.

Finally, another process consumes the session data from the silver layer to update the gold layer. One job aggregates user activity to update **`customer_segment`** tables for marketing analytics. A separate, more frequent job calculates updated product recommendation scores for each active user and writes them to a low-latency database. This gold layer table directly serves the recommendation engine on Fusion-Mart's website, providing personalized and timely suggestions.

Fusion-Mart can rapidly experiment with new recommendation algorithms in the gold layer while the integrity of the conformed session data in the silver layer remains guaranteed. A feature flag in the application's configuration serves as the assurance artifact, allowing analysts to safely switch between old and new recommendation models for A/B testing without disrupting the user experience. Therefore, a raw clickstream event is systematically refined through the bronze, silver, and gold layers to become an actionable insight that powers a personalized customer experience.

Cautions and limits

While the medallion architecture provides significant benefits for data quality and governance, its implementation comes with challenges that require careful management. Organizations must address potential drawbacks related to cost, latency, and operational overhead to ensure a successful deployment.

One primary concern is increased storage cost and data duplication, as data exists in at least three forms. This can be mitigated by implementing data lifecycle policies to transition older bronze data to cheaper, archival storage tiers and by leveraging platform features like zero-copy cloning for development environments. Another limitation is processing latency; the multi-hop nature of the architecture is well-suited for batch and near real-time use cases, but may not meet the demands of applications requiring sub-second latency. For such cases, a governed exception or a separate streaming path may be necessary. Finally, the architecture introduces governance overhead. Managing schemas, access controls, and quality checks across three layers demands robust tooling and a dedicated data governance strategy to prevent the layers from degrading into a disorganized data swamp.

Organizations can manage storage costs with intelligent tiering even as data volumes grow, ensuring long-term archival without excessive expense. An automated cost-monitoring alert that triggers when a project's storage budget is projected to be exceeded acts as a practical assurance mechanism. Therefore, adopters must proactively manage the inherent challenges of storage costs, processing latency, and governance complexity to realize the architecture's full value.

Glossary

Key terms are defined to ensure a common understanding of the concepts discussed in this section:

- **Medallion architecture**: A data design pattern that organizes data into three progressive layers, such as bronze (raw), silver (cleansed and conformed), and gold (curated and business-ready), to systematically improve data quality and reliability.
- **Immutable data**: Data that, once written to storage, cannot be altered or deleted. This principle is fundamental to the bronze layer, guaranteeing an auditable and replayable history of all source data.
- **Data lake**: A centralized repository for storing vast quantities of structured, semi-structured, and unstructured data. In this architecture, the data lake typically hosts the bronze and silver layers.

Expert notes

For experienced practitioners, implementing a medallion architecture at scale involves leveraging advanced tools and concepts to enhance efficiency, governance, and security. These notes offer deeper insights for building a mature and robust data platform.

Adopting **Open Table Formats** (**OTF**) like Apache Iceberg, Delta Lake, or Hudi is critical for modern implementations. These formats bring database-like features such as ACID transactions, schema evolution, and time travel (data versioning) directly to data lakes, which are essential for managing the silver and gold layers reliably. A comparison of these formats can be found in *Table 4.3*. Furthermore, establishing formal data contracts, agreements between data producers and consumers that define schema, semantics, and quality metrics, can prevent breaking changes and ensure stability as the platform evolves. Finally, security should be enforced with programmatic, fine-grained controls. Instead of granting broad access, each data processing job should operate with a least-privilege **identity and access management** (**IAM**) policy, granting permissions only to the specific source and destination paths required for its task.

Advanced teams can implement atomic transactions on their data lake using OTF, ensuring data consistency without sacrificing the scalability of cloud object storage. A pre-commit hook in the data pipeline's source code repository that validates data contracts before code can be merged serves as a powerful assurance artifact to enforce governance automatically. Therefore, mature implementations enhance the medallion pattern with OTF for reliability, data contracts for governance, and least privilege IAM policies for security.

The following table shows the comparison of OTF for multi-cloud portability:

Feature	Delta	Iceberg	Hudi	Notes or cloud caveats
ACID transaction model	Supported	Supported	Supported	Ensures data integrity across clouds.
Metadata and catalog integration	Supported	Supported	Supported	Decouples data from physical storage.
Schema evolution	Supported	Supported	Supported	Enables non-disruptive data contract changes.
Partition evolution	N/A	N/A	N/A	Prose does not contain this information.
Time travel (Versioning)	Supported	Supported	Supported	Critical for data audit and version rollbacks.
Row-level operations (UPDATE/DELETE/ MERGE)	Supported	Supported	Supported	Enables reliable concurrent data modifications
Compute engine support	Strong Spark integration	Engine-agnostic (Trino, Flink)	N/A	Ecosystem fit is a key selection criterion

***Table 4.1**: Comparison of OTF for multi-cloud portability*

Related topics

This section focuses on medallion architecture principles and layer design. For related topics addressed elsewhere in this book:

- ETL/ELT tooling and implementation, see *Chapter 5, Building Scalable Data Pipelines*
- Data mesh organizational considerations, see *Chapter 2, Understanding Data Mesh, Lakehouse, and Medallion*

Multi-cloud deployment

A multi-cloud strategy is a foundational component of modern enterprise architecture, enabling organizations like Fusion-Mart to enhance resilience and strategic flexibility. By distributing workloads across multiple cloud providers, a business can select best-of-breed services, optimize costs, and mitigate the significant risks of vendor lock-in. This approach is critical for maintaining business continuity, as an outage with one provider does not cripple operations. This strategy allows an enterprise to innovate using diverse cloud capabilities, even as it maintains centralized governance over its data assets. A version-controlled configuration file for the identity federation trust policy acts as a concrete assurance artifact for this setup.

Key considerations

Successfully implementing a multi-cloud architecture requires careful planning around data structure, security, and operational trade-offs. The following points are critical for designing a robust and manageable solution:

- **Mapping data architecture to cloud services**: A logical data framework like the medallion architecture must be mapped to concrete services in each cloud. For Fusion-Mart, this could mean landing raw data in the bronze layer using cost-effective storage like AWS S3, while performing transformations for the gold layer in a powerful analytics engine like GCP BigQuery. This allows the architecture to leverage the unique strengths of each provider while maintaining a consistent data lifecycle.
- **Designing secure cross-cloud access**: Data access between clouds must be secure and governed by a centralized control plane that manages authentication and authorization. Using a mechanism like workload identity federation allows an application in one cloud to securely and temporarily assume a role in another, eliminating the need for risky long-lived credentials. This approach ensures that an organization can build a resilient data platform across clouds and still maintain a unified security posture.
- **Evaluating implementation trade-offs**: A multi-cloud strategy introduces complexities that must be managed. These include financial costs from data egress charges, security risks from an expanded attack surface, and data integrity challenges due to varying consistency models across services. Understanding these frictions, as outlined in *Table*

4.1, is essential for making informed decisions that balance flexibility with operational stability. A regularly audited IAM role with a feature flag to disable cross-cloud access provides a clear rollback path.

Therefore, a successful multi-cloud strategy depends on mapping logical data layers to optimal cloud services, securing cross-cloud access via identity federation, and proactively managing the inherent trade-offs in cost, security, and consistency.

The following table lists the common multi-cloud frictions and mitigation strategies:

Area	Typical pitfall	Detection signal	Mitigation	Residual risk
Data egress cost	High data transfer costs between cloud providers	Monitoring data transfer volume and costs	Optimize data transfer, use regional services, and negotiate with providers.	Some level of cost is unavoidable for hybrid models.
Auth or identity federation	Misconfigured access or long-lived credentials	IAM access reviews, audit logs for STS	Use workload identity federation, least-privilege policies, and temp credentials	Configuration errors can still grant excessive access.
Network latency	Performance degradation for cross-cloud apps	**Application performance monitoring** (**APM**)	Use private interconnects (e.g., cloud interconnect), co-locate services	Inherent latency between different cloud regions.
Skill or tooling fragmentation	Increased operational overhead, required expertise	Slower project delivery, high training costs	Use cloud-agnostic tools (IaC), central governance, and cross-training	Some platform-specific knowledge is always required.
Observability gaps	Difficulty troubleshooting cross-cloud issues	High **mean-time-to-resolution** (**MTTR**), siloed logs	Centralized logging and monitoring platform, standardized metrics	Complex integration, potential for blind spots
Cross-cloud policy drift	Inconsistent security policies create gaps	Security audits, unexpected access alerts	**Policy-as-code** (**PaC**), version-controlled configurations, central repo	Manual break-glass changes may not be reconciled

Table 4.2: *Common multi-cloud frictions and mitigation strategies*

Process or steps

Establishing secure cross-cloud data access follows a clear sequence of identity federation, permission granting, and data retrieval:

1. **Establish trust**: An application in a source cloud (e.g., GCP) requires data from a target cloud (e.g., AWS). To initiate this, you configure Workload Identity Federation, creating a trust relationship that allows the GCP service account to assume a specific AWS IAM role. This action results in the GCP application gaining the ability to request temporary, short-lived credentials from the AWS **Security Token Service** (**STS**).
2. **Grant permissions**: With trust established, the assumed AWS role needs permissions to access the specific resource. You must attach a least-privilege IAM policy to the role, granting only the necessary permissions, such as `s3:GetObject` on a specific S3 bucket. This ensures the application's access is strictly limited to its required function, preventing unauthorized data exposure or modification.
3. **Access data securely**: The application now uses the temporary AWS credentials to make an API call to the target service. This process enables seamless data access across cloud boundaries without embedding long-lived static credentials in applications. The data is then transferred, ideally over a private network connection, from the target (AWS) to the source (GCP), fulfilling the request as depicted in the generic path in *Figure 4.2*. The audit trail from the STS provides a concrete record of every credential request and role assumption.

Therefore, the process for secure cross-cloud access involves federating identities to establish trust, applying least-privilege policies to grant permissions, and then using temporary credentials to retrieve data over a secure channel.

Concrete example

Consider a practical scenario for Fusion-Mart where a data analytics job running on a Spark cluster in GCP needs to process daily sales transaction logs stored in an AWS S3 bucket. Instead of moving terabytes of data permanently, the Spark job can read it directly and securely. The GCP service account associated with the Spark cluster is configured to trust an AWS IAM role through workload identity federation. When the job runs, it automatically acquires temporary AWS credentials from the STS. Using these credentials, it accesses the S3 bucket over a private cloud interconnect link, ensuring data remains off the public internet. Fusion-Mart can run its analytics workloads on GCP's specialized hardware and still access its historical data archives stored cost-effectively in AWS. A Terraform or CloudFormation template defining the entire cross-cloud IAM relationship serves as a version-controlled, auditable artifact of the security configuration.

Therefore, a Fusion-Mart Spark job in GCP can securely process data in an AWS S3 bucket by using workload identity federation to obtain temporary credentials and read the data over a private network connection.

Cautions and limits

While powerful, a multi-cloud strategy introduces challenges that require proactive management to prevent unforeseen costs and operational issues. Transferring data out of a cloud provider's network incurs egress charges, which can accumulate rapidly and significantly impact budgets if not monitored. Furthermore, in multi-cloud architectures with cross-region replication, data synchronization delays can occur, meaning a read operation in one region may not immediately see the results of a write in another region. This replication lag can cause data integrity problems in tightly coupled workflows. By proactively managing these challenges, an enterprise can achieve multi-cloud benefits without succumbing to common pitfalls like budget overruns or data integrity issues. It is also crucial to weigh the complexity of different resiliency patterns, as shown in *Table 4.2*, and to avoid anti-patterns like multi-primary writes across clouds without a sophisticated conflict resolution strategy. A summary of these common frictions and their mitigations is available in *Table 4.1*. A cloud cost management dashboard with configurable alerts for egress traffic provides a concrete assurance mechanism against unexpected bills.

Therefore, the key limitations to managing in a multi-cloud environment include controlling data egress costs, designing applications to handle replication lag in distributed systems, and avoiding overly complex resiliency patterns that increase risk.

The following table shows the failover patterns vs. complexity:

Pattern	RPO/RTO	Ops complexity	Cost	Notes
Read-through mirror	hrs/hrs	Low	$	Cold standby for BI
Active or passive zone	min/hrs	Med	$$	Writer single, reader multi
Active or active multi-region	min/min	High	$$$	Conflict and governance-heavy

***Table 4.3**: Failover patterns vs. complexity*

Glossary

The following are the key terms that were discussed in this section:

- **Workload identity federation**: A security mechanism that allows a workload in one environment (like a GCP VM) to assume an identity in another environment (like an AWS IAM role) without using long-term credentials.
- **Identity federation**: A system that links a person's or service's digital identity and attributes across multiple security domains. It enables a centralized **Identity Provider (IdP)** to authenticate a principal once, allowing it to access resources in different cloud environments.

- **Egress charges**: Fees charged by cloud providers for transferring data out of their network to the public internet or another cloud.
- **Eventual consistency**: A consistency model in distributed systems where, if no new updates are made to a given data item, all accesses to that item will eventually return the last updated value.
- **Multi-cloud strategy**: An approach where an organization uses services from more than one public cloud provider to meet different technical or business requirements.
- **Control plane**: The part of a system that manages configuration, policy, and access control. In this context, it handles authentication and authorization for data access.
- **Data plane**: The part of a system that handles the actual movement of data packets based on the rules set by the control plane.

Expert notes

For organizations seeking to maximize data portability and minimize vendor lock-in, adopting OTF like Apache Iceberg or Delta Lake is a critical best practice. These formats decouple the data's structure and metadata from the underlying cloud storage, allowing the same dataset to be read by different query engines across multiple clouds, as compared in *Table 4.4*. This abstraction layer is fundamental to a truly portable data lakehouse. Additionally, always enforce least-privilege access by creating narrowly scoped IAM policies to ensure federated identities can only perform their intended functions. Advanced teams can build a truly portable data lakehouse architecture, even as they integrate deeply with the native security services of each cloud provider. A shared source code repository for data access policies, with mandatory peer review before deployment, serves as a concrete artifact for maintaining security standards.

Therefore, experts can enhance multi-cloud architectures by using ITF for data portability and enforcing strict, version-controlled, least-privilege IAM policies for cross-cloud interactions.

The following table shows OTF in multi-cloud (Delta / Iceberg / Hudi):

Aspect	Delta	Iceberg	Hudi	Notes
Transaction model	Log	Manifest	Timeline	Replication / DR impact
Branching / Time travel	Yes	Yes	Limited	Governance tie-ins

***Table 4.4**: OTF in multi-cloud*

Related topics

This section focuses on architectural patterns for secure cross-cloud data access. For related topics addressed elsewhere in this book:

- Cloud migration strategies, see *Chapter 9, Cloud Migration and Coexistence Strategies*
- Data governance and compliance, see *Chapter 6, Data Governance and Compliance*

IAM and logging strategies in multi-cloud

Building a trustworthy data platform requires embedding security into its core design rather than applying it as an afterthought. Foundational to this approach are robust IAM and comprehensive logging strategies. IAM acts as the gatekeeper, defining who can access which resources and perform specific actions, while logging provides a detailed, immutable record of all activity. Together, they form a critical security layer that enables organizations to enforce policies, detect threats, and prove compliance, ensuring data is both accessible to authorized users and protected from unauthorized use.

Key considerations

Before implementing specific security controls, it is essential to understand the guiding principles that underpin a secure data platform. These considerations form the strategic foundation for tactical decisions about tools and policies. A well-defined strategy allows an organization to scale its security posture effectively, even as the platform evolves, with a clear audit trail available through a version-controlled repository that tracks all policy changes.

- **Enforce the PoLP**: This fundamental security concept dictates that any user, service, or system component should only have the minimum permissions necessary to perform its intended function. Instead of granting broad access, you should define granular roles and policies that restrict actions to specific data sets or operations, which is crucial when mapping logical layers to physical services, as shown in *Figure 4.1*. This minimizes the potential damage from accidental misuse or a compromised account.
- **Establish centralized and immutable logging**: All components within the data platform, from storage and compute to networking services, must generate detailed logs of their activities. These logs should be aggregated into a central, secure repository where they are made immutable to prevent tampering. This consolidated audit trail is indispensable for security monitoring, incident response, and demonstrating regulatory compliance.
- **Integrate security controls for layered defense**: IAM and logging do not operate in isolation; they are part of a defense-in-depth strategy that includes network security and data protection. While network controls like private endpoints restrict traffic, IAM policies determine which authenticated identities can use those pathways. Similarly, while encryption protects data confidentiality, IAM controls who can manage the encryption keys and access the decrypted data.

Regulatory example

In financial services, **Sarbanes-Oxley (SOX)** compliance requires demonstrable controls over financial data. The centralized, immutable logs described here provide the audit trail that regulators inspect, helping demonstrate who accessed which data and when. Similarly, GDPR's right to be forgotten requires evidence that personal data was deleted; the audit logs help provide this documentation.

Therefore, a secure data platform is built on the principle of least privilege, supported by centralized, immutable logging, and integrated into a multi-layered defense strategy.

Process or steps

Implementing a secure access control policy involves a systematic process to ensure permissions are granted correctly and verifiably. This procedure ensures you can enable new data workflows while maintaining a strict security posture, with a feature flag allowing for the immediate rollback of any new policy that causes unintended side effects:

- **Define access requirements**: The process begins when a user or automated service requires access to a data resource. Clearly document the identity of the principal (the user or service), the specific resources they need to access (e.g., a specific table or storage path), and the exact actions they need to perform (e.g., read, write, delete).
- **Implement least privilege policy**: Based on the defined requirements, construct a granular IAM policy that grants only the specified permissions. Avoid using wildcards or overly broad permissions. This policy is then attached to a role, and the principal is assigned to that role, adhering to **role-based access control (RBAC)** best practices.
- **Audit and verify access**: After deploying the policy, use the centralized logging system to monitor the principal's activity. Query the access and audit logs to confirm that the principal can perform their required tasks and, just as importantly, that they are blocked from performing any unauthorized actions, thus verifying the policy's effectiveness.

Therefore, secure access is achieved by first defining specific needs, then implementing a granular, least-privilege IAM policy, and finally verifying its enforcement through centralized audit logs.

Concrete example

A practical scenario illustrates how these security principles function together. Imagine a data analyst who needs read-only access to the `customer_insights` table within the gold layer of the data platform to build a sales dashboard. The platform is built on cloud services, mapping logical constructs to physical infrastructure as detailed in *Figure 4.1*. To grant this access securely, the platform administrator does not assign permissions directly to the analyst's user account. Instead, they leverage an existing `DataAnalyst` role configured with RBAC. This role has an attached IAM policy that explicitly allows read-only actions on all tables within the gold layer's database schema. The analyst is added to the `DataAnalyst` group, which inherits this role. When the analyst queries the `customer_insights` table, the platform's IAM service validates their role and permissions before granting access. Simultaneously, the database and storage services generate access logs, recording the analyst's identity, the query they ran, and the exact time of access. This log entry is shipped to the central logging service, providing a clear audit record that can be reviewed later. This entire process enables productivity without compromising security, and the access grant can be revoked instantly by removing the analyst from the group, as confirmed by an automated alert on the audit trail.

Therefore, assigning a data analyst to a pre-defined `DataAnalyst` role with read-only permissions to the gold layer provides necessary access while generating a complete audit trail of their activity.

Cautions and limits

Implementing robust IAM and logging is critical, but it comes with challenges that must be managed proactively. While these systems enhance security, they can also introduce complexity if not governed properly. Organizations can manage these challenges and still deliver value by using automated policy validation checks in their **continuous integration/continuous deployment** (**CI/CD**) pipeline to catch overly permissive rules before they reach production.

One significant risk is privilege creep, where users accumulate more permissions than they need over time, violating the principle of least privilege. This often happens as roles change or temporary access is not revoked. To mitigate this, you must implement regular access reviews and automated tooling to identify and prune unused permissions. Another challenge is managing the volume and cost of log data. Comprehensive logging can generate terabytes of data, leading to high storage costs and performance issues. A tiered storage strategy, where fresh logs are kept in hot, searchable storage and older logs are moved to cheaper, archival storage, is essential for cost management. Finally, in multi-cloud environments, managing identities and correlating logs across different providers can be complex, as noted in the cross-cloud read path shown in *Figure 4.2*. Using an identity federation service and a unified **security information and event management** (**SIEM**) system helps centralize control and visibility.

Additionally, policy validation rules are not stationary; regulatory requirements evolve, new threats emerge, and business needs change. Automated validation catches known violations, but periodic human review remains essential to update the rule set itself. Organizations should schedule quarterly policy reviews to ensure the automated checks reflect current requirements.

Therefore, to maintain a strong security posture, you must actively manage privilege creep through regular reviews, control log storage costs with a tiered strategy, and centralize identity management in multi-cloud settings.

Glossary

This section defines key terms used throughout the discussion of data platform security:

- **PoLP**: A security concept in which a user is given the minimum levels of access, or permissions, needed to perform their job functions.
- **RBAC**: A method of restricting network access based on the roles of individual users within an enterprise. Permissions are associated with roles, and users are assigned to roles.
- **IAM**: A framework of policies and technologies for ensuring that the proper people and services have the appropriate access to technology resources.
- **Private endpoint**: A network interface that provides secure and private connectivity to a cloud service using a private IP address from within a virtual network. It eliminates the need to expose data traffic to the public internet.
- **SIEM**: A solution that helps organizations detect, analyze, and respond to security threats before they harm business operations by aggregating and correlating log data from multiple sources.
- **Multi-factor authentication (MFA)**: A security process that requires users to provide two or more verification factors to gain access to a resource, such as a password and a code from a mobile app.

Expert notes

For teams with mature security practices, there are advanced techniques to further enhance the data platform's defenses. These methods allow for more dynamic and automated security controls, enabling faster responses to emerging threats even as the platform scales. A key assurance artifact here is the automated alert generated by a threat detection model, which can trigger a predefined incident response playbook.

Beyond standard RBAC, consider implementing **attribute-based access control** (**ABAC**). ABAC provides more dynamic and fine-grained control by granting permissions based on attributes of the user (e.g., department, location), the resource (e.g., data sensitivity level), and the environment (e.g., time of day). This is particularly useful for enforcing complex security requirements in large organizations. Furthermore, you should automate the lifecycle of IAM policies and access reviews using IaC tools like Terraform or CloudFormation. Storing IAM policies in a version-controlled repository provides a clear audit trail of all changes and enables automated validation to prevent misconfigurations. Finally, integrate your centralized logging platform with machine learning models to enable advanced threat detection. These models can analyze log patterns to identify anomalies, such as unusual data access patterns or attempts to escalate privileges, that might be missed by rule-based alerting systems.

Therefore, advanced security can be achieved by using ABAC for dynamic permissions, automating IAM with IaC, and applying machine learning to log data for proactive threat detection.

Related topics

This section focuses on IAM and centralized logging strategies for data platforms. For related topics addressed elsewhere in this book:

- Data privacy and compliance regulations, see *Chapter 6, Data Governance and Compliance*
- Network security in multi-cloud, see *Chapter 9, Cloud Migration and Coexistence Strategies*

Delta Lake and other lakehouse table formats for multi-cloud

A multi-cloud medallion architecture demands robust data interoperability and consistency to function effectively within strict security boundaries. OTFs like Delta Lake, Apache Iceberg, and Apache Hudi are foundational to this design. They provide a critical abstraction layer over cloud-native object storage, preventing vendor lock-in and enabling advanced data lake capabilities. By offering features such as ACID transactions, schema evolution, and time travel (data versioning), OTFs ensure data integrity and auditability, making them essential for a resilient, **sub-30-minute recovery** (**RB-30**) model across disparate cloud environments.

Key considerations

When designing a multi-cloud data platform, several key factors must be evaluated to ensure success. The following considerations are paramount:

- OTF is integral to creating a portable and interoperable medallion architecture. They decouple the data's logical structure from the underlying physical storage, allowing different compute engines in different clouds to read from and write to the same dataset. This enables consistent analytics and data processing across cloud boundaries while maintaining a unified governance posture. A data contract versioned in Git specifies the approved table format and schema evolution rules, ensuring all teams adhere to the standard.
- The choice between Delta Lake, Iceberg, and Hudi depends on specific workload requirements related to transactions, data versioning, and governance. Each format offers unique strengths in areas like write performance, concurrent access patterns, and ecosystem integration. Evaluating these capabilities is crucial for selecting the right technology for each medallion layer to optimize performance and still meet stringent compliance needs, as detailed in the comparison in *Table 4.4*.

- Securing data access between cloud providers is non-negotiable and requires a defense-in-depth strategy. This involves using federated identity for temporary credentials, enforcing private network paths to prevent data exposure to the public internet, and implementing strict access policies. This layered security model allows authorized workloads to access data efficiently without compromising the security and integrity of the entire data lake.

Therefore, selecting the right open table format and implementing a Zero Trust security model are foundational for building a resilient, interoperable, and secure multi-cloud medallion architecture.

Process or steps

Establishing a secure cross-cloud read path involves a sequence of steps to ensure authenticated, private, and auditable data access:

1. **Federate identity with OIDC**: To avoid static, long-lived credentials, establish a trust relationship between the IdP in the source cloud and the IAM service in the target cloud using **OpenID Connect** (**OIDC**). This allows a workload in the source cloud (e.g., a Kubernetes pod) to request a short-lived security token from its local IdP. This token can then be exchanged for temporary credentials in the target cloud, enabling secure access without exposing secrets.

2. **Establish private network connectivity**: To prevent data from traversing the public internet, create a private network link between the VPCs in each cloud. For cross-cloud connectivity, use technologies like VPN tunnels, AWS Direct Connect with Google Cloud Interconnect, or third-party SD-WAN solutions. For intra-cloud private connectivity, use AWS PrivateLink or Google Cloud Private Service Connect. This action ensures that all data transfer occurs over a secure, low-latency backbone, isolating sensitive data from external threats and reducing egress costs.

3. **Configure least-privilege access**: In the target cloud, create a specific IAM role that the source workload can assume. The role's trust policy must be scoped to the OIDC provider and specific identities (e.g., a particular service account). The permissions policy should grant only the necessary actions (e.g., `s3:GetObject`) on specific resources (e.g., a single S3 bucket prefix). This allows secure data access and still prevents lateral movement or unauthorized data manipulation. An automated pipeline job checks for and revokes any IAM roles with trust policies that are too permissive.

4. **Enable granular audit logging**: To ensure full visibility into data access, enable data event logging on the object storage service in the target cloud (e.g., AWS CloudTrail Data Events). This action captures every object-level API call, such as `GetObject` and `PutObject`, providing a detailed audit trail. The resulting logs are indispensable for security monitoring, incident response, and demonstrating compliance. This entire flow is visualized in *Figure 4.3*.

Therefore, a secure cross-cloud process combines OIDC-based identity federation for temporary credentials, private networking to isolate traffic, least-privilege IAM roles for access control, and data-level event logging for comprehensive auditing.

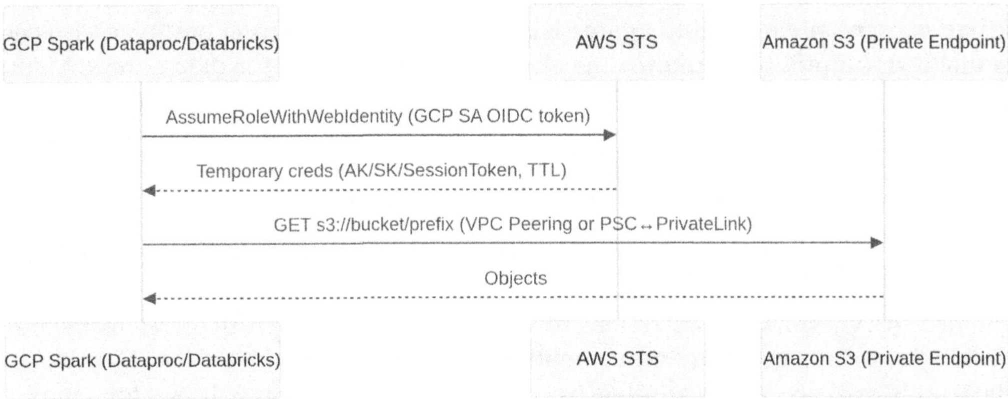

Figure 4.3*: Cross-cloud data access flow using open table formats and federated identity*

Concrete example

A common multi-cloud scenario involves an analytics job running on a **Google Kubernetes Engine** (**GKE**) cluster in GCP that needs to query a gold medallion table stored in an AWS S3 bucket. The table is managed by Delta Lake. The process begins by configuring GCP's workload identity federation to trust the GKE service account's identity provider. In AWS, an IAM role is created with a trust policy that allows principals federated through the specific GCP identity pool to assume it. The role's permission policy grants read-only access to the target S3 bucket.

When the Spark job on GKE starts, it uses its native service account token to call the AWS **Security Token Service** (**STS**), presenting the token as proof of identity. AWS STS validates the token with GCP's OIDC provider and, upon success, returns temporary AWS credentials to the Spark job. The job then uses these credentials to access the Delta Lake table files in S3. All network traffic is routed through a secure connection established between the GCP VPC and AWS VPC via a dedicated VPN tunnel or cloud interconnect, bypassing the public internet. The query engine accesses data directly and still adheres to the target account's security posture. A CloudTrail log entry in AWS confirms the **`GetObject`** API call was made using the temporary role assumed via OIDC, providing a clear audit record of the access.

Therefore, a GCP Spark job can securely query data in AWS S3 by using workload identity federation to obtain temporary AWS credentials and routing all traffic over a private network connection.

Cautions and limits

While OTF and cross-cloud patterns provide significant flexibility, they also introduce complexities that must be managed. Network latency between clouds can impact query performance, especially for highly interactive workloads. Data transfer costs can also become substantial if not carefully monitored, as cloud providers charge for data egress. Managing identity and permissions across multiple cloud IAM systems adds operational overhead and requires meticulous configuration to prevent security gaps. Furthermore, ensuring compatibility between different versions of compute engines (e.g., Spark, Trino) and OTF libraries across clouds is critical to avoid runtime errors or data corruption.

To mitigate these challenges, teams should design data pipelines to minimize cross-cloud chatter, performing as much processing as possible within a single cloud before transferring only the necessary results. Organizations can manage these complexities even as the number of cloud environments grows by implementing robust monitoring for network performance and data transfer costs. A feature flag in the deployment pipeline allows for temporarily disabling cross-cloud replication during a version upgrade to prevent data inconsistencies. Adopting IaC for deploying IAM policies and network connections helps ensure consistency and reduces the risk of manual configuration errors.

Therefore, multi-cloud data architectures face challenges in network latency, egress costs, and identity management, which can be mitigated through careful pipeline design, cost monitoring, and automated infrastructure deployment.

Glossary

The following are the key terms that have been discussed in this section:

- **OTF**: A specification for managing large-scale analytical datasets in data lakes. It defines how table metadata is organized, enabling features like ACID transactions, schema evolution, and time travel on top of object stores.
- **ACID transactions**: A set of properties that guarantee data validity despite errors, power failures, and other mishaps. In the context of data lakes, it enables reliable, concurrent data modifications.
- **Time travel**: The capability to query a data table as it existed at a specific version number or timestamp. It is primarily used for auditing, reproducing experiments, and safely rolling back data to a known-good state.
- **Workload identity federation**: A feature that allows workloads running outside a cloud provider (e.g., on-premises or in another cloud) to access resources securely without needing to manage long-lived service account keys.
- **OIDC**: An identity layer built on top of the OAuth 2.0 framework. It allows clients to verify the identity of an end-user based on the authentication performed by an authorization server.

- **Virtual private cloud (VPC) peering**: A networking connection between two VPCs that enables you to route traffic between them using private IP addresses. This allows resources in different VPCs to communicate as if they were within the same network.

Expert notes

When selecting an open table format for a multi-cloud architecture, experienced practitioners should look beyond the core features. The choice between Delta Lake, Iceberg, and Hudi often comes down to the specific read/write patterns and the maturity of the ecosystem in each cloud. For instance, Delta Lake's deep integration with Databricks and Spark makes it a strong choice for Spark-heavy environments, while Iceberg's engine-agnostic design provides flexibility for organizations using a diverse set of query engines like Trino or Flink. A detailed feature breakdown is available in *Table 4.3* and *Table 4.4*.

For performance optimization, consider data layout strategies such as partitioning and Z-ordering (or equivalent data skipping techniques in Iceberg/Hudi) based on common query predicates. This is especially important in a cross-cloud scenario, as minimizing the amount of data scanned directly reduces network latency and egress costs. Engineers can optimize query performance while minimizing costly cross-cloud data egress by using predicate pushdown aggressively. A monthly cost report, tagged by workload, provides a clear audit trail of data transfer expenses, allowing for continuous refinement of data access patterns and partitioning schemes.

Therefore, advanced implementation requires choosing an OTF based on ecosystem fit and workload patterns, while optimizing performance and cost through intelligent data layout and predicate pushdown.

Related topics

This section focuses on OTF patterns for multi-cloud medallion platforms. For related topics addressed elsewhere in this book:

- Data pipeline performance optimization, see *Chapter 5, Building Scalable Data Pipelines*
- Cloud networking strategies, see *Chapter 9, Cloud Migration and Coexistence Strategies*

Artificial intelligence pipeline integration

A modern data platform's value is realized when it delivers trusted, high-quality data to downstream consumers like AI and ML models. For a regulated enterprise, a well-architected data foundation is a strategic imperative, supplying the pristine data required by sensitive applications such as fraud detection and customer personalization. These models are highly susceptible to data quality issues, where minor inconsistencies can lead to erroneous predictions and significant business impact. Therefore, integrating AI pipelines with the core data platform requires a robust, secure, and auditable framework to ensure a continuous flow of high-fidelity data, forming a verifiable data supply chain.

Key considerations

Before implementing a data access pattern for AI pipelines, it is essential to understand the foundational principles that ensure success. These considerations address data quality, security posture in complex environments, and the specific controls needed to protect data assets:

- A well-architected data foundation, such as the medallion architecture, is crucial for supplying high-quality data to AI models. This layered approach systematically refines data from its raw state (bronze) to cleansed (silver) and finally to business-ready aggregates (gold), ensuring that models train on reliable, consistent, and trusted information. This structured refinement is critical for applications like fraud detection, where data integrity directly impacts model accuracy and business outcomes.

- AI pipeline integration in multi-cloud, regulated environments demands robust, secure, and auditable data access patterns. As organizations use services across multiple cloud providers, they must establish a consistent security framework to manage complexity and meet compliance mandates. A standardized pattern ensures that security and auditability are built-in, not bolted on, providing a defensible posture regardless of where data or compute resides.

- Implementing key security measures is fundamental to protecting data consumed by AI systems. Enforcing the use of private endpoints isolates data traffic from the public internet, using short-lived credentials minimizes the risk associated with compromised keys, and applying minimal **access control lists** (**ACLs**) adheres to the principle of least privilege. These controls work together to create a secure data path from the storage layer to the AI service, even as new models are deployed, and can be managed via a central feature flag system to enforce adoption.

Therefore, a successful AI data integration strategy relies on a quality-focused data architecture, a standardized and auditable access pattern for multi-cloud environments, and strict, foundational security controls.

Process or steps

Creating a secure and auditable data supply chain for AI follows a repeatable pattern that ensures every data access event is attributable, authorized, protected, and verifiable:

1. **Establish identity**: The process begins when a need for data access is identified. An automated workflow assigns a distinct, granular machine identity, such as an IAM role or a service account, to each AI component, including training jobs and inference endpoints. This action results in every subsequent request being clearly attributable to a specific, known principle, eliminating anonymous access.

2. **Define policy**: With an identity established, the system requires explicit authorization to proceed. A narrowly scoped access policy is attached to the identity, granting the minimum permissions necessary, such as read-only access to specific gold layer tables. The result is that the identity is governed by the principle of least privilege, preventing unauthorized actions or access to non-essential data.
3. **Enforce secure path**: When the AI service attempts to retrieve data, the request must traverse a secure network path. The architecture enforces this by routing all traffic through private endpoints, completely isolating it from the public internet, a process visualized in *Figure 4.2*. This action results in a confidential and integrity-protected data flow, mitigating risks of interception or man-in-the-middle attacks.
4. **Generate logs**: As the data is accessed, the cloud platform's native services must record the event. Services like AWS CloudTrail or Google Cloud Audit Logs are configured to capture every detail of the access attempt, including the principal, resource, timestamp, and outcome. The result is a comprehensive, immutable log of all data access activities.
5. **Produce evidence**: To meet compliance requirements, the recorded logs must provide verifiable proof of control effectiveness. The logs are aggregated into a central SIEM system, where they can be correlated and analyzed. This final step produces a complete, tamper-evident evidence trail that demonstrates to auditors that the data supply chain is secure and compliant, and an automated script can confirm daily log ingestion to ensure the completeness of this audit trail.

Therefore, the identity | policy | secure read path | logs | evidence pattern provides a structured, end-to-end process for ensuring that all AI data access is authenticated, authorized, secure, and auditable.

Concrete example

Consider Fusion-Mart, a regulated enterprise that uses a multi-cloud strategy to power its AI-driven fraud detection system. The system's training pipeline, running in Google Cloud's Vertex AI, needs to access transactional data stored in the gold layer of their medallion architecture, which resides in an Amazon S3 bucket. This architecture maps logical data layers to specific cloud services, as illustrated in *Figure 4.1*. To achieve this securely, Fusion-Mart implements the five-step access pattern:

1. First, the Vertex AI pipeline is assigned a Google Cloud service account, establishing its identity. This identity is then federated to assume an AWS IAM role using workload identity federation.
2. Second, a least-privilege IAM policy is attached to this role, granting it read-only access to the specific transactions table in the S3 bucket.
3. Third, the data retrieval occurs over a secure read path, using Google Private Service Connect and AWS PrivateLink to ensure the connection between GCP and AWS never traverses the public internet.

4. Fourth, both Google Cloud Audit Logs and AWS CloudTrail logs capture the entire transaction, from the initial request in GCP to the S3 object access in AWS.

5. Finally, these logs are streamed to a centralized SIEM, creating a unified evidence trail. This allows security teams to run a single query to trace the data access event across both clouds, providing a complete audit record that can be automatically packaged into a quarterly compliance report. This allows Fusion-Mart to innovate using best-of-breed cloud services while still maintaining stringent security and auditability.

Therefore, Fusion-Mart's multi-cloud fraud detection system exemplifies the secure access pattern by using federated identities, least-privilege policies, private networking, and centralized logging to create an auditable data trail from a GCP AI service to data in AWS S3.

Cautions and limits

Implementing a secure, multi-cloud data access pattern introduces manageable challenges that require careful planning. The complexity of managing identities, policies, and private networks across different cloud providers can lead to misconfigurations if not handled systematically. For instance, maintaining consistent IAM policies or ensuring private network routes are correctly configured between clouds can become an operational burden, and these common multi-cloud frictions are detailed in *Table 4.1*. Furthermore, this pattern presupposes a certain level of organizational maturity, including established cloud governance practices and expertise in both networking and security.

To mitigate these risks, organizations should use IaC with tools like Terraform to define and manage identities, policies, and network resources in a repeatable and auditable manner. This approach allows teams to develop and deploy cloud infrastructure even as security policies are centrally enforced through a pre-commit hook that validates resource configurations. Adopting policy-as-code frameworks and leveraging automated security posture management tools can further reduce the risk of human error and ensure continuous compliance. By addressing these challenges proactively, an organization can build a robust and scalable multi-cloud architecture.

Therefore, the primary challenges of this pattern are operational complexity and the need for mature governance; these can be mitigated through extensive automation, IaC, and automated policy validation.

Glossary

The following are the key terms that have been discussed in this section:

- **Medallion architecture**: A data architecture pattern that organizes data into bronze (raw), silver (cleansed), and gold (aggregated) layers to progressively improve data quality and usability for analytics and AI.

- **IAM**: A security framework of policies and technologies for ensuring that the proper entities (users and services) have the appropriate access to technology resources.
- **Private endpoint**: A network interface that uses a private IP address from a virtual network, enabling a private and secure connection to a cloud service without exposing traffic to the public internet.
- **SIEM**: A solution that aggregates and analyzes log data from various sources to help organizations detect, analyze, and respond to security threats in near real time.

Expert notes

For advanced implementations, organizations can enhance security and efficiency beyond the baseline pattern. Instead of using static credentials, services should dynamically request short-lived credentials from an STS. This practice significantly reduces the risk of credential leakage, as the tokens expire automatically after a brief period. Furthermore, the adoption of OTF like Apache Iceberg or Delta Lake can greatly simplify cross-cloud data access, as shown in *Figure 4.3*. These formats decouple the data from a specific query engine, allowing different compute services across clouds to operate on the same dataset seamlessly, with a comparison of options available in *Table 4.3*.

Automating the entire evidence-gathering process is another key area for experts. By integrating security tools directly into the CI/CD pipeline, teams can validate infrastructure and policy code against compliance rules before deployment. This proactive approach allows developers to innovate quickly and still adhere to strict security controls. The CI/CD pipeline can include a validation stage that automatically scans for and blocks overly permissive policies, ensuring security is maintained without manual intervention.

Therefore, advanced implementations should leverage short-lived credentials for enhanced security, use OTF to enable cross-cloud data portability, and embed automated compliance checks directly into CI/CD pipelines.

Related topics

This section focuses on secure data access patterns for AI/ML pipelines. For related topics addressed elsewhere in this book:

- AI model deployment and monitoring, see *Chapter 7, MLOps for AI Model Deployment and Monitoring*
- Data privacy regulations, see *Chapter 6, Data Governance and Compliance*

Lab: Medallion consistency check

Now that you have learned the principles of layered refinement, this lab puts theory into practice by simulating the tiny medallion consistency check. You will work as a data engineer for Fusion-Mart, ensuring that core tables (e.g., customers) maintain structural consistency, in terms of row counts, key coverage, and schema evolution, as they flow from the raw bronze layer to the curated gold layer. The lab evaluates whether the medallion configuration is reliable and trustworthy enough for downstream analytics.

Lab environment: All labs run in GitHub Codespaces with a dev container. This lab produces `labs/ch04/artifacts/result.json` containing status, checks, metrics, and messages, ensuring consistent governance evidence regardless of technology.

> Note: **The lab uses accept/reject for simplicity. In the book text, we use Go/Retry/Kill to describe the same decision workflow.**

Conclusion

This chapter equipped you with the essential skills to design a concrete medallion architecture blueprint, specifically tailored to operate within boundary gate constraints. You have learned to map medallion layers to specific cloud services for the Fusion-Mart case study, design secure cross-cloud data access patterns, and critically evaluate the trade-offs of multi-cloud implementation choices. This practical foundation ensures your data architectures are robust, secure, and adaptable across diverse cloud environments.

Having established a solid architectural foundation, the next crucial step is bringing your data to life. The next chapter will guide you through constructing these pipelines, leveraging powerful tools to automate and manage your data flow from ingestion to transformation. You will move from design principles to hands-on implementation, ensuring your medallion architecture is populated with high-quality, accessible data.

Fusion-Mart's multi-cloud medallion architecture now provides consistent data quality across AWS, Azure, and GCP, enabling global operations with unified governance.

CHAPTER 5
Building Scalable Data Pipelines

Introduction

If a deployment fails at 2 AM, can your team restore service before business hours? This chapter answers the second question: *How do we guarantee production safety?* by building pipelines that pass the RB-30 gate.

Ensuring the resilience and rapid recovery of data pipelines is paramount in modern data architecture. A critical benchmark for this reliability is the RB-30 gate: the requirement to restore a pipeline to a known good operational state within 30 minutes of detecting an incident. This is not about reverting to a 30-minute-old snapshot, but about having a proven, rehearsed recovery process that returns the system to full, accurate operation within that time window. The RB-30 gate is a critical principle of the Four-Gate Governance Model introduced in the *Reading Guide,* which we will detail in this chapter. The single change highway pattern you will learn in this chapter is based on architectures and practical constraints found in real-world environments with the most stringent requirements, such as finance, IoT, and e-commerce.

This chapter guides you through constructing robust and scalable data pipelines designed with this very challenge in mind. We will leverage industry-standard tools, such as Kafka for real-time data streaming, a **data build tool** (**dbt**) for data transformation, and Terraform for **infrastructure as code** (**IaC**), to build a system that is not only efficient but also inherently reversible. By the end of this chapter, you will be proficient in building a complete data pipeline using Kafka, dbt, and Terraform. More importantly, you will learn to execute and document a timed rollback rehearsal, proving your pipeline's resilience and readiness.

Finally, you will learn to generate the lab's **`result.json`**, the evidence that records your measured recovery time. This evidence demonstrates your system's operational integrity, which will be automatically recorded into the audit trail, a single auditable ledger.

Structure

This chapter covers the following topics:

- Streaming vs. batch ingestion
- Data transformation and testing with dbt

- Infrastructure as code with Terraform
- Pipeline performance and cost optimization
- Delta Lake integration for versioning and time travel
- Industry case studies
- Lab: Mini single change highway

Objectives

By the end of this chapter, you will be able to build scalable pipelines using Kafka for streaming ingestion, dbt for transformations, and Terraform for infrastructure deployments and choose between streaming and batch ingestion based on data characteristics, latency requirements, and cost constraints. You will design resilient pipelines that survive infrastructure failures and service disruptions while continuing to operate without data loss, while you implement the RB-30 gate to guarantee rollback within 30 minutes, proving this capability through rehearsals before production deployment. Finally, you will deploy complete, reversible pipelines from ingestion through transformation with automated rollback capabilities that protect production systems.

Streaming vs. batch ingestion

Choosing the right data ingestion method is a critical architectural decision that directly impacts a data pipeline's performance, reliability, and recoverability. This choice between processing data continuously (streaming) or in scheduled groups (batch) must be grounded in specific use case requirements, such as latency tolerance and throughput needs. For any system to be resilient, it must handle data processing failures and retries gracefully. This is achieved through idempotent design, where operations can be repeated without creating duplicate data or errors. A well-designed ingestion strategy ensures that even as new data flows in, the system can recover from failures without data corruption, a capability confirmed through a documented and rehearsed rollback plan.

Key considerations

Selecting an appropriate ingestion strategy requires balancing business needs with technical realities. The classic 3Vs framework, volume, velocity, and veracity, helps frame this decision: how much data, how fast it arrives, and how trustworthy it is. The following points are central to designing a robust and recoverable data pipeline:

- **Streaming vs. batch trade-offs:** The primary decision between streaming and batch ingestion hinges on latency requirements. Streaming ingestion provides near real-time data for operational use cases like inventory management, while batch processing offers high throughput and cost-efficiency for analytical workloads like weekly sales reporting. You can deliver timely insights with streaming and still manage costs by reserving batch processing for less time-sensitive, large-scale analyses. In practice, many architectures are hybrid: a streaming layer handles real-time events while periodically enriching them with batch-processed reference data. For example, real-time clickstream events may be joined with nightly-updated customer segments. An automated alerting system that monitors data freshness against defined objectives provides assurance that business needs are being met.

- **Idempotency as a cornerstone:** Data pipelines must be built to withstand failures, which often cause data to be re-processed. Idempotency ensures that reapplying an operation multiple times produces the same result as applying it once, preventing data duplication or corruption. This is typically achieved in the data sink using `UPSERT` or `MERGE` operations based on a unique business key. This allows the system to advance with new data without compromising integrity, with correctness verified by an audit trail logging all data write operations.

- **Designing for replayability:** A resilient pipeline must support the ability to reprocess data to correct errors or recover from system failures. This requires immutable raw data storage (a bronze layer) and well-defined replay strategies, such as full or point-in-time reprocessing. The interaction between a **change data capture** (**CDC**) source and the data sink's behavior is a crucial factor in enabling effective replays. A version-controlled script that can trigger a data replay provides a concrete mechanism for recovery.

Therefore, the choice between streaming and batch ingestion must be driven by latency needs, but both methods require a foundational design based on idempotency and planned replayability to ensure data integrity and resilience, which is the basis for achieving the RB-30 gate standard.

Process or steps

Creating a resilient, idempotent ingestion process involves three key steps:

1. **Define a composite unique key:** A failure to uniquely identify each data event causes duplicates upon replay. To prevent this, you must define a robust, unique identifier for each record by combining multiple attributes. This results in a composite key that provides practical uniqueness, typically consisting of the record's natural business key (e.g., `transaction_id`), the source `event_time` timestamp, and a cryptographic hash of the record's content.

2. **Implement idempotent sink logic:** The arrival of duplicate records from upstream retries can corrupt the target dataset. To address this, implement `UPSERT` or `MERGE` logic in the data sink (e.g., the data warehouse table). This action uses the composite unique key to either insert a new record or update an existing one, resulting in a system where data can be re-delivered safely without creating duplicates.

3. **Establish and document replay procedures:** The potential for data corruption or system failure necessitates a clear recovery plan. To mitigate this risk, you must document the exact procedures for replaying data from your immutable source. This results in a reliable, auditable recovery mechanism that allows you to restore the system to a known good state quickly, with confidence in the process confirmed by a log file from a successful, timed rollback drill.

Therefore, achieve idempotent ingestion by first creating a composite unique key from the natural key, event time, and content hash, then using that key to drive **`UPSERT`** logic in the sink, and finally documenting the entire process for reliable, auditable recovery to support the RB-30 objective.

Concrete example

Consider Fusion-Mart's real-time inventory system, which processes events for product sales and returns. When a product with **`product_sku 'X-47A'`** is sold, the source database generates a CDC event. To process this idempotently, the pipeline constructs a composite unique key. The **`natural_key`** is the **`product_sku 'X-47A'`**. The **`event_time`** is the precise timestamp at which the sale occurred, for instance, **`'2023-10-26T10:00:05Z'`**. Finally, a SHA256 **`content_hash`** is computed from the event's payload, which includes details like **`quantity_change: -1`**.

The resulting composite key uniquely identifies this specific transaction. The data is then written to an inventory table in the data warehouse using a **`MERGE`** statement. If a network issue causes the streaming processor to send the same event twice, the **`MERGE`** operation, keyed on the composite identifier, will simply reapply the same update without inserting a duplicate row. This allows the system to process a continuous stream of inventory changes even as transient failures occur. Assurance is provided by a feature flag that can enable a 'dry run' mode, which logs intended changes without applying them, allowing for verification before production deployment.

Therefore, for a Fusion-Mart inventory update, combining the **`product_sku`**, **`event_time`**, and a **`content_hash`** creates a unique key that enables a **`MERGE`** operation to safely handle duplicate events without corrupting inventory counts.

Cautions and limits

While idempotent ingestion provides strong guarantees, engineers must be aware of potential challenges and limitations. In streaming systems, out-of-order or late-arriving data can cause incorrect state calculations if not handled properly. This can be mitigated by using event-time processing and watermarking techniques to correctly window and process data based on when it occurred, not when it arrived. For batch systems, the primary limitation is data staleness; the data is only as fresh as the last batch run. This can be managed by aligning the batch schedule with business requirements for data freshness, as defined in pipeline **service level objectives** (**SLOs**).

You can make progress on complex streaming logic while still ensuring stability by isolating development work. A concrete assurance artifact is an automated monitoring alert that triggers when the time difference between **event_time** and processing time exceeds a predefined threshold, indicating a potential data freshness issue.

Therefore, key limitations include handling late data in streaming systems, which requires event-time processing, and managing data staleness in batch systems, which is addressed by aligning batch frequency with business needs. For use cases requiring a complete history of record changes, such as tracking customer address updates over time, consider implementing **slowly changing dimension type 2** (**SCD type 2**) patterns. While this chapter focuses on idempotent **UPSERT** logic for current-state tables, SCD Type 2 preserves historical versions and is an industry-standard approach covered in data warehousing literature.

Glossary

The following are the key terms discussed in this section:

- **CDC:** A process that identifies and captures changes made to data in a database (inserts, updates, deletes) and delivers those changes to downstream systems.
- **Checkpointing:** The practice of periodically saving the state of a data processing job to durable storage, allowing it to resume from the last saved point after a failure.
- **Event-time:** The timestamp embedded in a data record that indicates when the event actually occurred at its source, as opposed to when it was processed.
- **Exactly-once processing (EOP):** A data processing guarantee ensuring that each record is processed one time and one time only, even in the presence of system failures.
- **Idempotency:** A property of an operation, meaning it can be applied multiple times without changing the result beyond the initial application.
- **Natural key:** A column or set of columns in a data table that uniquely identifies a record based on real-world, business-domain attributes (e.g., a user's email address).
- **UPSERT:** A database operation that will **UPDATE** an existing row if a specified value already exists in a table, and **INSERT** a new row if the value does not exist.

Expert notes

For experienced practitioners, optimizing idempotent pipelines involves more than basic **UPSERT** logic. When reprocessing large historical datasets, full table scans for **MERGE** operations can become performance bottlenecks. To mitigate this, partition target tables by date and limit the **MERGE** operation to only the relevant partitions affected by the replay. This allows you to rebuild a segment of history without impacting the entire table.

Another advanced technique involves managing schema evolution during replays. If a data schema has changed over time, replay logic must be able to handle historical versions. Using a schema registry to manage and version schemas ensures that reprocessing jobs can correctly interpret older data formats.

Therefore, experts optimize for resilience by using partitioning to limit the scope of `MERGE` operations during replays and implementing a schema registry to manage schema evolution, ensuring historical data can be processed correctly for rapid recovery.

> Caution: **Apply date-based partitioning only at high data volumes. Premature partitioning on small tables creates excessively small files, degrading performance. Additionally, be aware of partition skew: if most records fall into recent partitions (e.g., 2024 to 2025) while historical partitions (e.g., 2020 to 2023) remain sparse, query performance and maintenance operations become uneven.**

Related topics

This section focuses on the principles and patterns for designing idempotent data ingestion. For related topics addressed elsewhere in this book:

- Data governance and compliance, see *Chapter 6, Data Governance and Compliance*
- CI/CD pipeline configuration, see *Chapter 8, DevOps and CI/CD for Data Engineering*

Data transformation and testing with dbt

This section uses dbt to illustrate data contract enforcement and automated testing principles. For streaming-first architectures or complex CDC workflows, code-based approaches (PySpark, Scala) or managed services (Delta Live Tables, Auto Loader) may be more appropriate. The principles of codified contracts, automated validation, and timed rollback rehearsals apply regardless of the specific tool.

Building resilient data pipelines requires a shift from merely processing data to engineering for reliability. The central measure of this reliability is the ability to recover swiftly from failure. This article outlines a disciplined approach to constructing a data pipeline using tools like Kafka, dbt, and Terraform, with a primary objective: demonstrating a full rollback to a known good state in under 30 minutes. This capability is not an accident; it is the result of intentional design, incorporating data contracts, automated quality enforcement, and rigorous, documented rehearsals. Achieving this standard ensures data trustworthiness and operational stability.

Key considerations

Before building the pipeline, it is crucial to understand the core principles that enable rapid recovery. These considerations form the foundation for a resilient and maintainable system:

- **Design for recoverability:** The primary goal is to establish a recovery time objective of less than 30 minutes. This forces design decisions that prioritize rapid restoration over other concerns. You can build a complex pipeline and still ensure it is recoverable by embedding rollback mechanisms directly into the architecture. A version-controlled data store, like Delta Lake, provides a concrete assurance artifact by enabling point-in-time restores.
- **Codify everything:** The entire data pipeline, from infrastructure to transformation logic, must be managed as code. Using Terraform for infrastructure ensures environments are reproducible, while dbt codifies business logic, tests, and documentation. This approach allows you to advance development quickly without sacrificing stability. Codifying a data contract in dbt's `schema.yml` creates an executable specification for data quality.
- **Rehearse for reality:** A rollback plan is purely theoretical until it is tested under realistic conditions. A timed rehearsal is not optional; it is a mandatory exercise to validate the process, tooling, and documentation. Executing these drills regularly allows teams to improve their response time even as the system evolves. The documented outcome of a successful rehearsal serves as a verifiable record of the system's recovery capability.

Therefore, a sub-30-minute recovery objective is achieved by designing for recoverability, codifying all system components, and validating the process through timed rehearsals.

Process or steps

Building a data pipeline capable of a sub-30-minute rollback involves four key steps:

1. **Provision IaC:** Begin by defining all necessary cloud resources, such as storage buckets, data processing services, and permissions, using Terraform. This practice ensures that your environment is version-controlled, repeatable, and can be provisioned or torn down consistently. You can deploy new infrastructure confidently without introducing manual configuration errors. An example of defining data infrastructure with Terraform, providing an audit trail of all environmental changes.
2. **Implement data contracts and transformations:** With the infrastructure in place, use dbt to develop your data transformation models. Crucially, define explicit data contracts for your models using dbt's `schema.yml` files. These contracts specify column types, nullability constraints, and other quality checks, formalizing the expected shape and integrity of the data. This mapping of contract fields to dbt tests allows you to build features while ensuring data quality standards are met, as detailed in *Table 5.1*.
3. **Automate quality enforcement in CI/CD:** Integrate your dbt project into a CI/CD pipeline. This pipeline should automatically run your dbt tests on every proposed change, programmatically enforcing the data contracts. This automated workflow prevents code that produces low-quality data from being merged into the main branch.

A feature flag can be used to control the deployment of new models, allowing for a quick rollback if an issue is discovered post-deployment.

4. **Execute and document a timed rollback rehearsal:** Once the pipeline is operational, simulate a failure. Introduce a known data error and execute your documented rollback procedure. This typically involves using a feature like Delta Lake's time travel to revert a table to a previous version. Time the entire process from detection to resolution and record the final duration, as the recovery time in a central location for tracking and auditing purposes.

Therefore, the process involves defining data contracts and transformations in dbt, automating quality checks in a CI/CD pipeline, and validating the entire system with a timed rollback rehearsal, ensuring the pipeline meets the resilience requirements of the RB-30 gate. The following table lists the mapping data contract fields to dbt tests:

Contract field	Required?	Test(s)	Severity	Example
Column name and type	Yes	Native type check	`error`	**`user_id`** is **`integer`**
Nullability	Yes	**`not_null`**	`error`	**`order_id`** must not be null
Uniqueness	Yes (for keys)	**`unique`**	`error`	Each **`order_id`** is unique
Referential integrity	Yes (for FKs)	**`relationships`**	`error`	**`customer_id`** exists in **`customers`**
Value constraints	Often	**`accepted_ values`**	`warn` or `error`	**`status`** in (**'shipped'**, **'pending'**)
Source freshness	Yes	**`source freshness`**	`warn` or `error`	Source data is < 24 hours old
Custom business logic	As needed	Custom SQL query	`error`	**`revenue`** must be >= 0

Table 5.1: *Mapping data contract fields to dbt tests*

Concrete example

Imagine an e-commerce company's analytics team is responsible for a critical **`daily_sales`** report. A developer accidentally introduces a bug in a dbt model that causes the `revenue` column to be calculated incorrectly, inflating the numbers. The automated quality checks fail, but due to a misconfiguration, the build is not blocked, and the erroneous data reaches the production dashboard. An analyst quickly spots the anomaly and triggers an incident. The on-call data engineer consults the team's rollback runbook, which provides step-by-step instructions. They can proceed with the fix while the system remains available to most users.

The engineer uses a versioned data storage system to revert the **`fct_sales`** table to a timestamp immediately preceding the faulty deployment. After confirming the data is restored, they re-run the downstream dbt models to refresh the **`daily_sales`** report. The entire process, from the initial alert to the dashboard displaying correct data, takes 24 minutes. This outcome is logged, and the recovery time of 1440 seconds (24 minutes) is recorded in the audit trail, providing a clear record of the successful recovery.

Therefore, when a bug corrupted a sales report, the team used a runbook and versioned data storage to revert to a known good state and restore correct data in 24 minutes, documenting the time as proof of their recovery capability.

Cautions and limits

Achieving rapid rollback capability requires acknowledging and managing several inherent challenges. While you can build a robust recovery process, you must do so without ignoring potential failure points. One primary limitation is the dependency on upstream data quality; your pipeline can validate and transform data, but it cannot fix fundamental errors at the source. To manage this, implement source freshness tests in dbt to halt execution if source data is stale, which provides a clear signal of an upstream problem. Another challenge is the complexity of stateful systems outside the data warehouse, as rolling back a table does not automatically revert state in dependent applications. The mitigation is to define clear boundaries for your rollback procedure and use loosely coupled architectures. Finally, an outdated runbook is a significant risk during a real incident. This can be managed by treating documentation as code, storing it in version control, and requiring updates as part of the change review process, ensuring the runbook is always synchronized with the production system.

Therefore, be aware of limitations like upstream data quality, the complexity of stateful systems, and outdated documentation, and mitigate them with source freshness tests, clear boundaries, and treating runbooks as code.

Glossary

The following are the key terms discussed in this section:

- **dbt:** A command-line tool that helps data teams transform data in their warehouse using SQL. It handles dependency management, testing, and documentation.
- **Data contract:** An agreement between data producers and consumers on the expected structure, semantics, and quality of a data asset. In dbt, this is often defined in **`schema.yml`** files.
- **Terraform:** An IaC tool used to define and provision data center infrastructure using a declarative configuration language.
- **Rollback:** The process of restoring a system or data asset to a previously known good state, typically after a failure or the deployment of faulty code.

- **CI/CD:** A software development practice that automates the integration, testing, and deployment of code changes, ensuring a reliable and efficient release process.

Expert notes

For teams seeking to enhance their recovery capabilities, several advanced practices are worth considering. Implementing a staged promotion model separates experimental work from the production deployment path, allowing for innovation without compromising stability. This structure provides a safe environment to test significant changes before they are considered for production. Furthermore, ensure your dbt models are idempotent, meaning they can be run multiple times without changing the result beyond the initial run. This property is critical for safe and predictable recovery operations. For rollback rehearsals, you can achieve deterministic results by pinning dependencies to a known good state. Finally, to create a verifiable and machine-readable record of your recovery capability, log the results of each rehearsal using a structured schema. This structured log serves as a powerful audit trail for compliance and continuous improvement.

Therefore, advanced techniques include using a staged promotion model for safety, ensuring dbt models are idempotent, and creating a deterministic audit trail of timed rehearsals with structured logging.

Related topics

This section focuses on building a data pipeline with demonstrable sub-30-minute rollback capability. For related topics addressed elsewhere in this book:

- Data governance frameworks and PII management, see *Chapter 6, Data Governance and Compliance*
- CI/CD orchestration patterns, see *Chapter 8, DevOps and CI/CD for Data Engineering*

Infrastructure as code with Terraform

IaC provides a foundational shift for managing data systems, enabling organizations to provision and configure resources through code rather than manual processes. This approach is critical for building agile and resilient data pipelines. By defining infrastructure in version-controlled files, teams can automate deployments, ensure consistency across environments, and rapidly recover from failures. This allows teams to accelerate development cycles while maintaining system stability. The core principle is to treat infrastructure with the same discipline as application software, which is essential for meeting stringent operational objectives and ensuring that every change is testable, repeatable, and reversible, backed by a version-controlled known good pin file that defines a stable state, a process clearly visualized in the change flow in *Figure 5.1*.

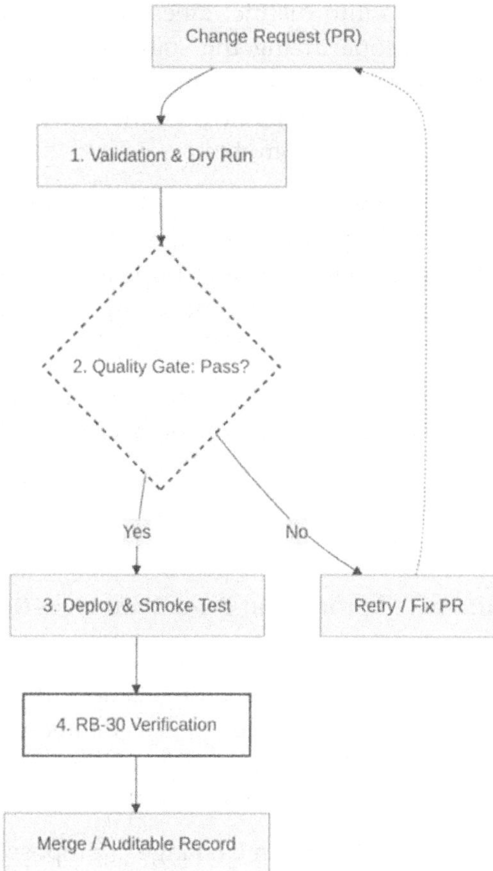

***Figure 5.1**: Minimal CI/CD flow*

Key considerations

Effective implementation of automated infrastructure and data pipeline management requires attention to several core principles. These considerations ensure that speed does not compromise stability and that governance is an enabler, not a bottleneck. The following points are central to building a robust and safe promotion process:

- **IaC as the source of truth:** Defining all infrastructure components, from networks to data services, in declarative code is the bedrock of this approach. Using tools like Terraform ensures that every environment is reproducible and auditable. This practice eliminates configuration drift and provides a clear, versioned history of all changes, which is crucial for debugging and rapid rollbacks.
- **Automated validation for safe promotion:** Integrating automated checks directly into the deployment pipeline is essential for maintaining quality and security. This

model separates workflows into distinct lanes, allowing for rapid innovation in an experiment lane without jeopardizing the stability of the production environment. This separation ensures that complex changes are validated in staged environments. Changes must pass a series of automated validations, such as security scans and data contract tests, before they can be promoted.

- **Proactive guardrails for pipeline integrity:** Beyond infrastructure, the data itself requires protection through operational guardrails. This includes enforcing data contracts to prevent breaking changes, and implementing monitoring to detect both schema and distribution drift. Designing for idempotent writes ensures that data can be reprocessed safely, which is fundamental for fault tolerance and recovery.

Therefore, a successful strategy relies on version-controlled infrastructure code, a staged promotion model with automated validation, and proactive data guardrails to ensure both agility and stability.

Process or steps

Implementing a safe and automated promotion process involves three key steps:

1. **Declare and version infrastructure:** The process begins when a developer defines the desired state of the infrastructure in a configuration file. By using an IaC tool like Terraform the configuration for resources such as databases and storage is committed to a version control system like Git. This action results in a versioned, auditable, and reproducible definition of the entire environment.
2. **Automate validation in CI/CD:** When a change is proposed via a pull request, a CI/CD pipeline is automatically triggered. This pipeline executes a series of validation checks, such as static code analysis, security scans, and data contract adherence tests, which are visualized in the minimal CI/CD flow. The result is an objective, automated assessment that confirms the change is safe to merge and deploy, allowing developers to push changes frequently and still trust that automated checks will catch breaking changes.
3. **Deploy and monitor with reversibility:** Once all checks pass, the change is automatically deployed to the target environment. The system's observability tools, such as metrics, logs, and traces, continuously monitor the health and performance of the newly deployed components. This provides immediate feedback on the impact of the change and ensures that if an issue is detected, the system can be quickly reverted to a previous known good state using the versioned IaC definition, with the process generating an auditable deployment log as proof of successful validation.

Therefore, the process involves declaring infrastructure as code, validating changes through an automated CI/CD pipeline, and deploying with robust monitoring to ensure safe and reversible promotions.

Concrete example

A data analytics team needs to add a new **`customer_segment`** column to a core data model to support a new marketing initiative. They begin by updating the dbt model definition and the corresponding data contract, marking the new column as nullable to ensure backward compatibility. Simultaneously, they modify the Terraform configuration to update the schema of the underlying data warehouse table. When they submit these changes in a pull request, the CI/CD pipeline automatically runs dbt tests to verify the data contract and executes a Terraform plan to preview the infrastructure changes. This allows the team to evolve the data schema to meet new business needs even as downstream consumers continue to operate without interruption. Following the staged promotion model, the change is first deployed to a staging environment for final validation before being promoted to production. A pull request with a complete history of code changes, review comments, and automated check results provides a clear audit trail for the modification.

Therefore, adding a new column illustrates how defining changes in code, validating them through an automated pipeline, and using a staged promotion model enables safe and auditable evolution of a data pipeline.

Cautions and limits

While powerful, adopting an IaC and automated governance model presents challenges that require careful management. The initial setup can be complex, demanding new skills in infrastructure management, testing, and pipeline automation. Organizations can adopt these powerful automation techniques while managing the inherent risks through disciplined processes and proactive monitoring. A primary risk is configuration drift, where manual changes made directly in the production environment cause it to diverge from the code-defined state; this can be mitigated by enforcing a strict code-only modification policy. Furthermore, the effectiveness of automated checks depends entirely on the quality and coverage of the tests. Inadequate testing can create a false sense of security, allowing defects to slip through. Finally, while IaC enables rapid rollbacks, the process is not always instantaneous and must be regularly rehearsed to ensure it functions as expected under pressure. Another risk is pipeline proliferation: the ease of spinning up new infrastructure with IaC can lead to dozens of redundant pipelines that could have been consolidated into simpler SQL or PySpark jobs. Maintain a clear inventory of deployed pipelines and regularly evaluate whether IaC-managed resources are still necessary. A scheduled, automated drift detection report provides assurance by regularly comparing the live infrastructure state against the version-controlled code.

Therefore, teams must manage the initial complexity, prevent configuration drift with strict policies, ensure comprehensive test coverage, and regularly practice rollback procedures to realize the full benefits of this model.

Glossary

Defining terms clearly helps teams communicate effectively while reducing ambiguity, with a shared, version-controlled glossary file acting as the assurance artifact:

- **Data contract:** A formal, machine-readable agreement between a data producer and its consumers that defines the schema, semantics, and quality expectations for a dataset.
- **Distribution drift:** A change in the statistical properties of data over time, such as shifts in the mean, variance, or cardinality of a column.
- **Idempotent:** An operation that can be applied multiple times without changing the result beyond the initial application. This is crucial for safely retrying failed pipeline jobs.
- **IaC:** The practice of managing and provisioning computing infrastructure through machine-readable definition files, rather than through physical hardware configuration or interactive configuration tools.
- **Observability:** The ability to measure the internal state of a system by examining its external outputs, such as logs, metrics, and traces.
- **Schema drift:** An unintended or undocumented change in the structure of data, such as adding or removing a column or altering a data type.

Expert notes

For experienced practitioners, maximizing the value of this operating model involves focusing on advanced techniques for resilience and efficiency. Designing idempotent write operations and replay-safe data transformations is critical for fault tolerance. This ensures that entire sections of a pipeline can be re-run without causing data duplication or corruption. Technologies like Delta Lake offer powerful features for this, such as time travel capabilities that allow for querying or restoring data to a specific point in time. Staged promotion also provides significant psychological safety, encouraging experimentation and innovation. Advanced teams can achieve high deployment velocity and still maintain stringent compliance by embedding evidence collection directly into their automated workflows. Retaining artifacts like a timed rehearsal log and a known good state pin transforms compliance from a manual chore into an automated, operational discipline. Using a feature flag system allows for the progressive rollout of new infrastructure, with an immediate rollback capability if monitoring detects any issues.

Therefore, experts should focus on designing for replay safety using tools like Delta Lake, leveraging the single change highway for innovation, and automating evidence collection to integrate compliance into operations.

Related topics

This section focuses on the principles and processes for safe, automated promotion of data infrastructure and pipelines. For related topics addressed elsewhere in this book:

- Machine learning model deployment, see *Chapter 7, MLOps for AI Model Deployment and Monitoring*
- CI/CD pipeline design, see *Chapter 8, DevOps and CI/CD for Data Engineering*

Pipeline performance and cost optimization

In data engineering, an architect's perspective is crucial for building economically viable systems. Performance, cost, and SLOs are not independent goals but coequal variables in a complex equation. Pursuing maximum performance without regard to budget or reliability targets is unsustainable. Similarly, aggressive cost-cutting that compromises performance can lead to missed SLOs and erode business value. Effective data pipeline optimization is a strategic exercise in balancing these interconnected forces to deliver reliable data products efficiently. Architects must balance these factors to build sustainable systems, delivering value while managing expenses. A key assurance artifact is a documented rollback plan, proven through timed rehearsals as tracked in recovery time logs.

Key considerations

An effective data pipeline strategy requires a holistic view of its operational and economic characteristics. The following points are central to this perspective:

- Performance, cost, and SLOs are fundamentally linked. Improving one aspect, such as reducing data processing latency, often requires a corresponding investment in resources, which increases cost. The SLO×cost trade-off model illustrates how achieving stricter SLOs, like higher availability or lower latency, necessitates higher expenditure. This allows teams to pursue higher performance targets and still remain within budget. The process is governed by a feature flag system that allows for immediate rollback of a new optimization if it proves too costly.
- FinOps provides a framework for managing these variables. By applying specific FinOps levers, teams can make informed decisions about resource allocation and usage. For example, choosing between batch and streaming processing or implementing auto-scaling mechanisms are distinct cost levers with different effects on performance and expenditure. This proactive management ensures that technical decisions align with financial constraints.
- SLOs must be specific, measurable, and tied to business outcomes. Vague goals are unhelpful; concrete targets provide clear benchmarks for success. Whether the goal is 99.9% data freshness within an hour or a query latency under 500 milliseconds, these

objectives guide optimization efforts and justify the associated costs by connecting them to tangible business value.

Therefore, optimizing data pipelines requires treating performance, cost, and SLOs as interconnected variables, managing them as critical constraints to ensure that recovery time remains within the RB-30 objective.

Process or steps

Optimizing a data pipeline is an iterative process of identifying inefficiencies, implementing targeted improvements, and measuring the results against defined objectives. The following steps provide a structured approach to this cycle:

- **Establish baselines and identify bottlenecks:** First, you must understand the current state of the pipeline. This involves collecting observability metrics on key performance indicators like latency, throughput, and resource utilization. By analyzing these metrics, you can pinpoint the specific stage, be it data ingestion, transformation, or loading, that is constraining overall performance or driving up costs. This data-driven diagnosis is the critical first step.
- **Select and implement an optimization strategy:** With a bottleneck identified, select an appropriate optimization technique. This could involve refactoring inefficient code, applying data partitioning to improve parallelism, or implementing a caching layer to reduce load on a database. The choice of strategy should align with the specific problem and consider the cost implications. This iterative cycle enables continuous improvement without disrupting production services. Each change is logged in an audit trail, providing a clear history of modifications for compliance and debugging.
- **Measure impact and validate against SLOs:** After deploying the change, measure the same metrics again to quantify the impact. Compare the new performance and cost figures against the established baselines and the target SLOs. If the optimization successfully improves performance and the pipeline still meets its SLOs within budget, the change is validated. If not, the process returns to the previous step to refine the strategy or try a different approach.

Therefore, a systematic optimization process involves establishing performance baselines, implementing targeted improvements like partitioning or caching, and validating the results against cost constraints and SLOs.

Concrete example

Consider Fusion-Mart's real-time personalization engine, which must deliver product recommendations with a latency under 50 milliseconds to meet its SLO. This system exemplifies the balance between performance, cost, and reliability. The pipeline ingests a high volume of user clickstream data using a streaming platform like Kafka and processes it with an in-memory computation engine to generate recommendations instantly.

Initially, the team identified that the recommendation model's feature lookup was a major bottleneck, causing latency to spike above the SLO during peak traffic. Following the optimization process, they implemented a caching layer using a high-speed in-memory database. This action drastically reduced lookup times by storing frequently accessed user profiles and product data closer to the computation engine. The team can deploy new recommendation algorithms even as the core service remains stable. A concrete assurance is the ability to restore the model registry to a previous known good state.

After deploying the cache, they measured latency and found it consistently stayed below 40 milliseconds. While this solution increased infrastructure costs due to the new caching servers, the business value of improved customer engagement and higher conversion rates justified the expense. The result was a faster, more reliable system that directly contributed to a better return on investment.

Therefore, Fusion-Mart met its sub-50ms recommendation latency SLO by implementing a caching layer, accepting a calculated cost increase to resolve a performance bottleneck and improve business outcomes.

Cautions and limits

While pipeline optimization is essential, it carries inherent risks and limitations that must be managed carefully. A primary caution is against premature or excessive optimization, which can introduce unnecessary complexity and make the system harder to maintain. Focusing on micro-optimizations that yield marginal gains can be a misallocation of engineering effort. It is more effective to target significant bottlenecks first.

Furthermore, every optimization strategy has limitations. For instance, materialized views can accelerate queries but increase storage costs and add write latency. Similarly, achieving exactly once semantics (a processing guarantee that each message is processed once and only once) is notoriously difficult and can impose significant performance overhead. Teams must weigh the benefits of such guarantees against their implementation and operational costs. Teams can cautiously introduce optimizations while having a safety net. An example artifact is a pinned, known good container image version for immediate redeployment. Finally, optimizations are not static; a change that improves performance today may become a bottleneck tomorrow as data volumes or query patterns evolve, requiring continuous monitoring and adaptation.

Therefore, avoid premature optimization that adds complexity, be aware that techniques like materialized views have cost and latency downsides, and recognize that optimizations require continuous re-evaluation as systems evolve.

Glossary

This section defines key terms used throughout the discussion.

- **SLO:** A specific, measurable target for a system's reliability, such as 99.9% uptime or a query response time of less than 200 milliseconds. SLOs are the formal benchmarks used to evaluate system performance.
- **Idempotency:** The property of an operation where executing it multiple times produces the same result as executing it once. This is critical in data pipelines for safely retrying failed operations without creating duplicate data.
- **Dead-letter queue (DLQ):** A dedicated queue that stores messages or events that a system failed to process after a certain number of retries. A DLQ prevents a single problematic message from halting the entire pipeline and allows for later analysis of the failure.
- **Exactly once semantics:** A data processing guarantee ensuring that each unique message is processed one time and one time only, preventing both data loss and data duplication. It is the strongest processing guarantee and is often complex to implement correctly.

Expert notes

For experienced practitioners, building resilient and efficient data pipelines involves implementing a minimal set of reliability patterns. A fundamental pattern is *retry with jitter*, which handles transient failures by re-attempting an operation after a randomized delay to avoid overwhelming a recovering service. When retries fail, a DLQ with a defined policy is essential to isolate problematic data without stopping the entire pipeline. This ensures the system can handle transient failures gracefully without losing data. A concrete assurance is the DLQ policy that archives failed messages for manual inspection.

To prevent system overloads, backpressure guardrails are crucial for allowing downstream services to signal when they are at capacity, thereby throttling the flow of data from upstream producers. For data integrity, designing for idempotency via natural keys ensures that reprocessing data does not create duplicates. This is a prerequisite for achieving the gold standard: exactly once semantics. Achieving this requires transactional writes from the processor, idempotent operations in the sink system, and robust state management. Finally, none of this is manageable without comprehensive observability metrics that track latency, throughput, and error rates, providing the necessary visibility to diagnose issues and verify that SLOs are met.

Therefore, advanced pipeline reliability depends on implementing patterns like retries with jitter, DLQs for error handling, backpressure for stability, and designing for idempotency to enable exactly once processing guarantees.

Related topics

This section focuses on the strategic balance of performance, cost, and SLOs in data pipeline optimization. For related topics addressed elsewhere in this book:

- CI/CD workflows for data projects, see *Chapter 8, DevOps and CI/CD for Data Engineering*
- Data governance and compliance, see *Chapter 6, Data Governance and Compliance*

Delta Lake integration for versioning and time travel

Data integrity and recoverability are foundational to reliable data pipelines. Delta Lake provides a powerful safety mechanism through its transaction log, which enables versioning and time travel capabilities. This allows teams to query or restore data to any previous state, transforming data lakes into auditable, enterprise-grade systems. Our operational philosophy centers on a deterministic repair strategy: we pin critical tables to a validated, known good version, ensuring that any recovery operation is predictable and swift. This allows teams to innovate on data models while still ensuring a rapid, deterministic recovery path, with the transaction log serving as a complete audit trail of every change.

Key considerations

Effective data recovery strategies using Delta Lake are built on several core principles. These considerations ensure that the technical capabilities of time travel translate into reliable operational practices:

- **Deterministic repair:** The cornerstone of a reliable recovery process is eliminating guesswork. By maintaining the known good pin file that stores a specific, validated Delta table version number, teams establish a definitive recovery target. This allows for aggressive data updates and still provides a safety net against errors, as recovery becomes a matter of executing a pre-defined command against a known state.
- **Explainability and auditability:** Every operation in Delta Lake creates a new table version and is recorded in the transaction log. This immutable history provides a complete audit trail, making it possible to trace data lineage and understand exactly how data has changed over time. This transparency is critical for debugging pipeline failures, satisfying compliance requirements, and building trust in the data.
- **Retention policy management:** The ability to time-travel is directly limited by the data retention period. This setting dictates how long historical data versions are kept before they become eligible for permanent deletion by the `VACUUM` command. A longer retention period provides a wider recovery window but increases storage costs, requiring a careful balance between operational safety, compliance needs, and budget.

Therefore, a successful Delta Lake recovery strategy depends on pinning to a known good version for deterministic repairs, leveraging the transaction log for full explainability, and carefully managing data retention policies to balance recovery windows with storage costs.

Process or steps

Executing a data rollback to a known good state involves a clear, sequential process:

1. **Identify incident and known good state:** The process begins when monitoring detects a data quality issue, such as from a failed downstream process or a direct data validation alert. The first action is to consult pipeline logs to pinpoint the time or version of the erroneous transaction. Concurrently, retrieve the last validated version number from the designated known good pin file.

2. **Execute rollback:** With the target version identified, the next step is to revert the table. This is accomplished using Delta Lake's `RESTORE TABLE` command, specifying the target version or a timestamp from before the incident occurred. This command atomically updates the table's metadata to point to the historical version, instantly making the known good data active without rewriting or moving large amounts of data.

3. **Validate and document:** After the `RESTORE` command completes, it is critical to validate the data's integrity by re-running data quality checks against the restored table. The entire incident, including start time, end time, actions taken, and the responsible engineer, must be documented. This structured approach enables rapid response even as the team investigates the root cause, with a feature flag that can be toggled to pause the problematic upstream pipeline, preventing further data corruption.

Therefore, the recovery process involves identifying the failure and the target known good version, executing the `RESTORE TABLE` command, and finally validating the data while documenting the event.

Concrete example

To illustrate the process, consider a scenario where a nightly batch job erroneously ingests malformed data into a critical `customer_orders` table. An automated data quality check fails and triggers an alert, initiating the recovery procedure. The on-call engineer begins a timed rehearsal to restore the table within the required 30-minute window. The team can restore service quickly without having to perform a complex and slow manual data correction, and the auditable record of the recovery event provides immutable proof of the process.

The recovery unfolds as follows:

- The engineer starts a timer and initiates the formal logging process, noting the start time and incident details in the auditable record.

- The engineer consults the version-controlled known good pin file and finds the last validated version is 125.
- The engineer executes the SQL command: **`RESTORE TABLE customer_orders TO VERSION AS OF 125;`**
- Once the command completes, they run a series of validation queries to confirm the data matches the expected state at version 125.
- With validation successful, the engineer stops the timer, calculates the total duration, and records the final **`recovery_seconds`** value (which must be 1800 or less) directly into the central audit trail as proof of compliance with the recovery objective.

Therefore, a real-world rollback involves an engineer using a known good version pin to execute a **`RESTORE`** command, validating the result, and documenting the entire timed process in a structured, auditable record to meet recovery time objectives.

Cautions and limits

While Delta Lake's time travel is a powerful feature, it is essential to understand its operational boundaries and potential risks. The most significant limitation is the **`VACUUM`** command, which permanently deletes data files that are no longer referenced by the transaction log and are older than the configured retention period. Once data has been vacuumed, it cannot be recovered via time travel. This makes the retention policy a critical configuration that must be set deliberately to ensure a sufficient recovery window for all data assets.

Another consideration is storage cost. Maintaining a long history of a large, frequently updated table can consume significant storage resources. Teams must balance the need for a long recovery window against budget constraints. Furthermore, while restoring data is straightforward, schema evolution can introduce complexity. If downstream applications have been updated to work with a newer table schema, rolling back to an older version with a different schema could cause application failures. Teams can manage storage costs effectively while preserving a sufficient window for forensic analysis and recovery, and a configuration alert that triggers when retention periods are set below a minimum safety threshold can prevent accidental data loss.

Therefore, a primary limitations of time travel are the finality of the **`VACUUM`** command and increased storage costs from long retention periods, with schema mismatches posing a potential challenge during recovery.

Glossary

Understanding these terms allows engineers to configure data systems for high performance and still maintain robust data protection, with a link to the official Delta Lake documentation for each term providing deeper context:

- **Delta Lake:** An open-source storage framework that brings ACID transactions to data lakes. It enables reliability, performance, and features like time travel on top of existing storage like S3 or **Azure Data Lake Storage** (**ADLS**).
- **Transaction log:** An ordered, atomic record of every transaction performed on a Delta table. This log is the single source of truth and is what enables reliable versioning and time travel.
- **Time travel:** The capability to query a Delta table as it existed at a specific version number or timestamp. It is used for auditing, reproducing experiments, and rolling back data to a known good state.
- **VACUUM:** A command used to clean up stale data files that are no longer referenced by a Delta table and are older than the specified retention threshold. This action is irreversible and is used to manage storage costs and permanently delete data.
- **OPTIMIZE:** A command that improves query performance by compacting many small files into a smaller number of large files. This reduces metadata overhead and improves I/O efficiency.

Expert notes

For teams looking to advance their operational maturity, several practices can enhance the reliability and efficiency of Delta Lake environments. One key area is the automation of recovery rehearsals. The entire process, from simulating a failure to executing the rollback and validating the result, can be scripted and integrated into a CI/CD pipeline. This ensures that recovery capabilities are continuously tested and validated without manual intervention. Automating these checks allows the data platform to scale efficiently without sacrificing operational stability, and a CI/CD pipeline job that automatically validates the known good pin file version before a rehearsal adds another layer of assurance.

Another advanced technique is to implement dynamic retention policies. Instead of a one-size-fits-all retention period, policies can be tailored based on data criticality, regulatory requirements, and update frequency. For instance, critical financial tables might have a 90-day retention period, while transient staging tables might have only a few days. Finally, for performance-critical workloads, consider using Z-ordering during `OPTIMIZE` operations. This technique co-locates related information in the same set of files, significantly improving query speed for filtered queries by reducing the amount of data that needs to be scanned. For teams using Databricks, **Delta Live Tables** (**DLT**), and Auto Loader, provide managed abstractions over these patterns. DLT simplifies pipeline orchestration with declarative quality expectations, while Auto Loader incrementally ingests new files with exactly-once guarantees. These tools are complementary to the principles discussed here and are detailed in Databricks documentation.

Therefore, advanced practices include automating recovery drills within CI/CD, implementing dynamic data retention policies based on table criticality, and using Z-ordering to optimize query performance.

Related topics

This section focuses on the operational mechanics of using Delta Lake's time travel for data recovery. For related topics addressed elsewhere in this book:

- Delta Lake transaction logs for compliance, see *Chapter 6, Data Governance and Compliance*
- Medallion architecture design, see *Chapter 4, Medallion Architecture in Multi-cloud*

Industry case studies

The true measure of the single change highway lies in its ability to mitigate mission-critical risks. This section moves beyond conceptual examples by illustrating how the RB-30 gate and its accompanying technical framework are essential for preserving financial integrity, ensuring regulatory compliance, and protecting physical safety across diverse industries.

Key considerations

Before examining the scenarios, it is crucial to understand the three primary failure modes that mandate the RB-30 constraint in a modern data platform:

- **Financial integrity risk:** Data failures that directly lead to immediate, quantifiable business loss (e.g., faulty pricing, incorrect inventory). The 30-minute recovery is essential to stop the bleeding.
- **Regulatory and audit violations:** Infrastructure or data quality failures that compromise the ability to meet legal compliance deadlines (e.g., transaction monitoring reports). RB-30 mitigates legal penalty risks.
- **Mission-critical safety:** Failures in processing data required for safety systems (e.g., filtering errors in sensor alerts). The system must guarantee immediate operational stability.

Mitigation framework

The mitigation of these mission-critical risks relies on the interplay of three core technical controls, which are applied sequentially within the single change highway:

- **Codified data contracts (Validation):** The pipeline must enforce quality rules and schema checks, preventing catastrophic data (e.g., bad sensor readings) from being processed. This is the first line of defense against the safety risk.
- **IaC:** The ability to quickly rebuild and redeploy the entire environment (Kafka, network, storage) is essential. This guarantees compliance recovery for the financial and regulatory risks.

- **Timed rollback rehearsal (Verification):** The final test proves the entire system can revert to a known good state. This validates the RB-30 promise.

Case studies

The following are the case studies:

- **E-commerce and pricing integrity:**
    - **Scenario**: A bug in a core dbt model corrupts Fusion-Mart's real-time inventory table, leading to overselling. The damage is quantified in millions per hour.
    - **RB-30 action**: The rapid RB-30 mechanism is triggered, leveraging Delta Lake's time-travel capability to instantly revert the inventory table to its pre-error state. The recovery is logged, proving the system halted the financial loss within the mandated time.
- **Financial services and regulatory compliance:**
    - **Scenario**: A buggy Terraform (IaC) script accidentally disables a production Kafka topic, halting real-time transaction monitoring needed for regulatory reporting.
    - **RB-30 action:** The RB-30 verification is executed on the IaC repository, proving the infrastructure can be rebuilt, and the Kafka service restored to a compliant state in under 30 minutes, mitigating severe regulatory penalties.
- **IoT and mission-critical safety:**
    - **Scenario**: A flawed data contract update is merged, causing the validation stage to incorrectly filter out essential sensor anomaly data before it reaches the alerting system.
    - **RB-30 action:** The single change highway's gate should block the change. If it fails to block, the RB-30 mechanism provides the final safety net, enabling the faulty logic to be rapidly reverted to restore the full integrity of the safety alerting pipeline.

Cautions and limits

This section discusses the risks associated with the RB-30 framework:

- **Initial overhead:** Implementing timed rehearsals and IaC requires significant initial investment and cultural adjustment.
- **Complexity mismatch:** The simplicity of the lab (single change highway) must not lead to the belief that real-world deployment is equally simple; complex infrastructure still requires discipline.

- **Outdated runbooks:** Rollback plans (runbooks) must be treated as code and continually tested and versioned alongside the pipeline to ensure they function under pressure.

Glossary

The context of this section requires a specific understanding of deployment and architecture terms:

- **Canary release:** A deployment strategy that exposes a new version of a service to a small subset of users first (Used in advanced mitigation).

Lab: Mini single change highway

In this lab, you will put theory into practice by simulating the mini single change highway, the fixed sequence of validation, gating, and deployment necessary to pass the RB-30 gate. Your objective is to run a simple governance check that proves your pipeline configuration has the right shape to support a safe, reversible change before committing to complex cloud infrastructure deployment.

Lab environment: All labs run in GitHub Codespaces with a dev container. This lab produces `labs/ch05/artifacts/result.json` containing status, checks, metrics, and messages, ensuring consistent governance evidence regardless of technology.

Note: **The lab uses accept/reject for simplicity. In the book text, we use Go/Retry/Kill to describe the same decision workflow.**

Conclusion

This chapter empowered you to construct a robust, scalable data pipeline using Kafka, dbt, and Terraform, culminating in the successful demonstration of an RB-30 gate pass. You now possess the practical skills to build such pipelines, execute and document critical rollback rehearsals, and meticulously record recovery time metrics for operational excellence. This hands-on experience ensures your data infrastructure is not only efficient but also resilient and recoverable.

Having mastered pipeline construction and rollback, our next focus shifts to ensuring the integrity and compliance of your data. The next chapter answers the third question: *How do we answer auditors?* by guiding you through implementing the audit trail. You will learn to leverage audit logs and data vault principles to maintain data quality, security, and regulatory adherence, building upon the scalable foundations established here.

Fusion-Mart's real-time inventory system now processes events with sub-50ms latency and proven 24-minute rollback capability.

CHAPTER 6
Data Governance and Compliance

Introduction

When an auditor asks *show me the evidence for this change*, can you answer in one click? This chapter answers the third question: *How do we answer auditors?* by building the audit trail.

This chapter implements the evidence and decision evaluation pattern from the Four-Gate Governance Model, a governance framework that evaluates changes through four sequential gates: boundary (project constraints), rb-30 (rollback capability), evidence (audit consolidation), and AI proposal (evaluating AI-generated proposals). While other chapters demonstrated boundary definition in *Chapter 2, Understanding Data Mesh, Lakehouse, and Medallion*, rollback validation in *Chapter 5, Building Scalable Data Pipelines*, and AI proposal evaluation in *Chapter 7, MLOps for AI Model Deployment and Monitoring*, this chapter focuses on consolidating all evidence into a unified audit trail that enables informed Go/Retry/Kill decisions.

Making informed decisions requires complete, verifiable evidence from all prior gates. Whether changes arrive from human teams, CI/CD pipelines, or AI-generated proposals, you need a unified audit trail that consolidates boundary compliance, rollback evidence, and validation results into a single, immutable source of truth.

This chapter establishes a robust framework for data governance and compliance, leveraging audit logs and data vault principles. We focus on building the audit trail, a comprehensive and immutable repository that serves as your single source of audit truth. When changes can arrive from multiple sources at increasing speed, scattered logs across different systems become unmanageable. The audit trail consolidates all decision logs, test results, and artifacts into a single, auditable URL.

This consolidated approach transforms evidence management from a compliance burden into a strategic asset. Auditors gain instant access to complete proof chains. Engineering teams make faster Go/Retry/Kill decisions with confidence, and your organization demonstrates regulatory compliance automatically, without manual evidence gathering for each audit.

Structure

This chapter covers the following topics:

- Data vault fundamentals for historical lineage
- Audit logging and secure log storage in multi-cloud
- Audit trail best practices for data governance
- Delta Lake transaction logs for compliance and auditing
- Navigating BFSI, healthcare, and retail regulatory mandates
- Lab: Data governance with a tiny data vault

Objectives

By the end of this chapter, you will be able to consolidate decision logs, test results, and artifacts into a single audit trail URL that provides complete evidence for compliance and implement WORM policies using S3 Object Lock to ensure evidence remains immutable and tamper-proof for required retention periods. You will apply data vault 2.0 principles to create auditable lineage that tracks how data and decisions evolve without destroying history, while you implement the evidence gate to satisfy regulatory requirements by producing complete, verifiable evidence for every platform change. Finally, you will design evidence workflows that enable development velocity while providing auditors the transparency they require for validation.

Data vault fundamentals for historical lineage

Passing a formal evidence and decision gate is a critical requirement for enterprise operations, demanding a complete and verifiable history of any significant change. The primary objective is to create a comprehensive and immutable audit trail URL that consolidates all decision logs, test results, and supporting artifacts into a single, tamper-proof location. This approach ensures that development teams can continue to deliver value and innovate rapidly, while still providing auditors with a transparent and reliable record of governance. By leveraging a WORM storage policy and data vault 2.0 modeling principles, organizations can establish a tamper-evident audit trail, where the audit trail URL serves as the definitive assurance artifact for compliance.

Key considerations

To successfully create an auditable audit trail, several core principles must be addressed:

- **Data immutability:** The foundation of a trustworthy audit trail is the guarantee that evidence, once recorded, cannot be altered or deleted. Implementing a WORM policy on a cloud storage service is essential for this purpose. This ensures that artifacts like logs and test results are preserved in their original state for the required retention period, allowing teams to deploy features confidently without compromising the integrity of the historical record. A direct link to the WORM-configured storage bucket serves as a powerful assurance artifact.

- **Evidence consolidation:** Auditors require a clear, centralized, and easily accessible path to all relevant information for a given change. A single, comprehensive audit trail URL prevents the need to hunt for scattered documents across multiple systems, which can lead to incomplete reviews and audit failures. Consolidating all required contents into one location streamlines the audit process and demonstrates organizational maturity. The consolidated URL itself becomes the primary assurance artifact presented to auditors.
- **Auditable data lineage:** It is not enough to store the evidence; you must also model its history and relationships in a way that is transparent and traceable. Data vault 2.0 principles provide the ideal framework for this, using its structure of Hubs, Links, and Satellites to create a non-destructive, append-only history of all changes, as depicted in *Figure 6.1*. This allows for rapid integration of new data sources and still provides a complete, point-in-time view of the past, with a query showing the full history of a change acting as the assurance artifact.

Therefore, a successful audit trail relies on WORM storage for immutability, a single URL for consolidated access, and data vault modeling for auditable data lineage.

The following figure shows the core components of a data vault model: Hubs, Links, and Satellites:

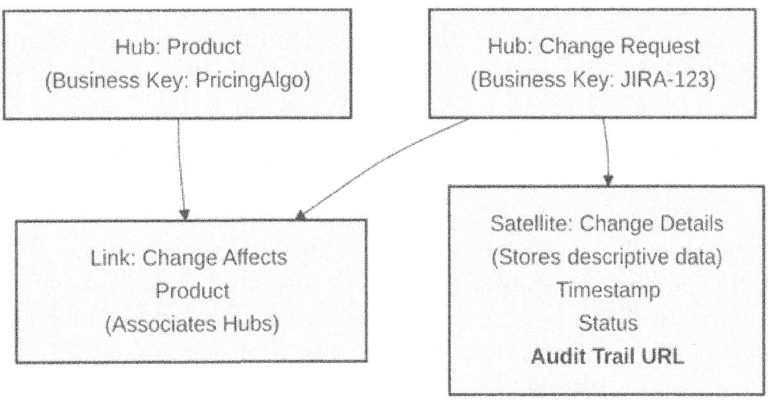

***Figure 6.1**: Core components of a data vault model*

Process or steps

Creating a compliant audit trail involves three key steps to ensure data is captured, secured, and presented in an auditable format:

1. **Consolidate artifacts:** The initial step is to gather all evidence related to a specific change into a single, structured package. This includes decision logs from project management tools, automated test results from **continuous integration** (**CI**) pipelines, deployment manifests, and any other supporting documentation specified in your evidence matrix, which can be modeled after *Table 6.1*. This action transforms scattered information into a complete and coherent evidence package, ready for archival.

2. **Implement WORM storage:** The consolidated package must be stored in a tamper-proof location. This involves uploading the package to a cloud storage bucket, such as Amazon **Simple Storage Service** (**S3**) or Google Cloud Storage, that has been configured with a WORM retention policy. This action ensures the evidence is immutable for a defined period, satisfying stringent audit requirements while allowing development to proceed without manual intervention in the archival process. A versioned deployment manifest within this package serves as a key assurance artifact.

3. **Model and load data lineage:** Finally, metadata about the change and its evidence must be loaded into a data warehouse modeled using data vault 2.0 principles. A Hub identifies the unique business entity (e.g., the change request), Links connect it to other entities (e.g., the system affected), and Satellites store descriptive attributes, including the immutable URL to the evidence package in WORM storage. This final action creates the auditable data lineage, as visualized in *Figure 6.1*, which allows anyone to trace a change from its inception through to its evidence.

Therefore, the process involves consolidating all artifacts into a package, securing it in WORM storage, and modeling its metadata in a data vault to create a complete audit trail.

Concrete example

Imagine a Fusion-Mart development team is deploying an update to a product's pricing algorithm. To pass the required audit, they follow the established process. First, they gather the **`Jira-123`** ticket detailing the change approval, a test summary file from their CI pipeline, and a deployment manifest. They organize these files into a directory named **`change-123`**. This consolidation ensures all necessary proof is in one place, allowing the team to move forward with deployment and still satisfy governance checks.

Next, the team's automated deployment script uploads the entire **`change-123`** directory to a pre-configured Amazon S3 bucket with Object Lock enabled. Upon successful upload, the script receives a permanent, versioned URL for the **`change-123`** package. This URL is the immutable audit trail link. The test summary report, accessible via this URL, serves as a concrete assurance artifact proving quality standards were met.

Finally, a data loading process populates the enterprise data vault. It creates an entry in the **`ChangeRequest_Hub`** for **`Jira-123`**, links it to the **`Product_Hub`** for the relevant product, and adds a record to the **`Evidence_Satellite`**. This satellite record contains metadata about the change and, most importantly, stores the immutable S3 URL. Now, an auditor can query the data vault for **`Jira-123`** and immediately retrieve a tamper-proof link to all associated evidence, providing a complete and verifiable history.

Therefore, a pricing algorithm update is audited by packaging its artifacts, storing them in a WORM-enabled S3 bucket, and recording the immutable URL in a data vault satellite.

Cautions and limits

While creating an audit trail is a powerful assurance mechanism, teams must be aware of potential challenges. One key consideration is the cost associated with long-term WORM storage; retaining large volumes of data can become expensive. To manage this, organizations should implement data lifecycle policies that automatically transition older, less critical evidence to more cost-effective storage tiers where compliance regulations permit. Another challenge is the inherent inflexibility of WORM storage; once data is written, it cannot be corrected. If erroneous evidence is uploaded, the mitigation is not to delete it but to upload a corrected version and update the data vault model to flag the original as superseded. This approach allows for rapid correction even as it preserves a full, auditable history of the error and its resolution. A pre-flight checklist that must be signed off on before committing artifacts to WORM storage is a practical assurance artifact to prevent such errors.

Furthermore, defining the scope of relevant artifacts can lead to scope creep, where teams are unsure what to include. This ambiguity can slow down the process and create audit risk. The solution is to establish a clear and explicit evidence matrix, similar to *Table 6.1*, that defines the required documentation for different types and risk levels of changes. This provides teams with clear guidelines, enabling them to proceed efficiently while ensuring all compliance requirements are met.

Therefore, manage the challenges of WORM storage by using lifecycle policies for cost, flagging data for corrections, and defining a clear evidence matrix to prevent scope creep.

Implementation trade-off

The audit trail approach requires upfront investment in tooling, automation, and team training. For agile consultancy projects with tight timelines, a lightweight alternative, such as a simple Git repository with tagged releases and README documentation, may provide sufficient auditability without the full WORM infrastructure. Product-based organizations with long-term compliance obligations will benefit most from the complete audit trail pattern. Evaluate your project's risk profile, regulatory requirements, and timeline before committing to the full implementation.

The following table lists the evidence: what/where/how long:

Evidence	Source	Example query or export	Retention
Access log	SIEM	`principalId='svc-ingest' AND action='GetObject'`	13 months
Delta txn log	_delta_log	`version BETWEEN X AND Y`	7 years

***Table 6.1**: Evidence consolidation process and audit artifacts*

Glossary

The following are the key terms that have been discussed in this section:

- **Data lineage:** A record of the data's journey, detailing its origin, transformations, and movement over time. It provides a verifiable trail for audit and analysis.
- **Data vault 2.0:** A data modeling methodology designed for enterprise data warehousing. It consists of Hubs, Links, and Satellites to enable scalable, flexible, and auditable historical data storage.
- **Hub:** A core component in a data vault model that contains a unique list of business keys (e.g., **`customer_id`**, **`product_sku`**). It represents a core business concept.
- **Link:** A data vault component that establishes a relationship or transaction between two or more Hubs. It represents the associations in business processes.
- **Satellite:** A data vault component that stores descriptive, time-variant attributes for a Hub or a Link. It holds the historical context and descriptive details.
- **Immutability:** The property of data or storage that prevents any modification or deletion after initial creation, ensuring evidence remains trustworthy and tamper-proof throughout its lifecycle.

Expert notes

For experienced practitioners, the audit trail concept can be enhanced with further automation and advanced techniques. The process of gathering and uploading artifacts should be fully integrated into the **continuous integration/continuous deployment** (**CI/CD**) pipeline. This automation ensures that evidence collection is consistent, timely, and not subject to human error, allowing development teams to accelerate delivery and still maintain a perfect audit trail. An automated pipeline job that archives build artifacts to the WORM location is a robust assurance artifact in this context.

Furthermore, the use of cryptographic hashes can elevate the integrity of the audit trail. When loading metadata into the data vault Satellite, include a hash (e.g., SHA-256) of the evidence package. This allows auditors to programmatically verify that the content in the WORM storage has not been tampered with since it was first archived, providing an even stronger layer of assurance.

Finally, the temporal nature of the data vault model, combined with technologies like Delta Lake, enables powerful time-travel queries, as shown in *Listing 6.6*. This allows auditors or support teams to reconstruct the exact state of the business or an application at any point in the past, offering unparalleled capabilities for forensic analysis and root cause investigation without compromising ongoing operations.

Therefore, experts can enhance the audit trail by automating artifact collection in CI/CD, using cryptographic hashes for verification, and leveraging time-travel queries for advanced forensic analysis.

Related topics

This section focuses on the strategic framework for creating an immutable audit trail using data vault principles and WORM storage. For related topics addressed elsewhere in this book:

- ETL/ELT pipeline implementation, see *Chapter 5, Building Scalable Data Pipelines*
- Cloud platform configuration, see *Chapter 9, Cloud Migration and Coexistence Strategies*

Audit logging and secure log storage in multi-cloud

In modern data operations, demonstrating accountability and integrity is essential for maintaining trust and navigating regulatory requirements. This is achieved through a foundational strategy combining WORM retention policies with a complete, end-to-end audit trail. This approach ensures every significant action is immutably recorded, enabling rapid incident response and forensic analysis without compromising the integrity of the evidence. By transforming system activity into durable, verifiable proof, as illustrated in the high-level flow of *Figure 6.2*, organizations can turn lengthy audit processes into efficient, evidence-based reviews, supported by concrete assurance artifacts like a verifiable WORM proof file.

The following figure shows the audit trail end-to-end:

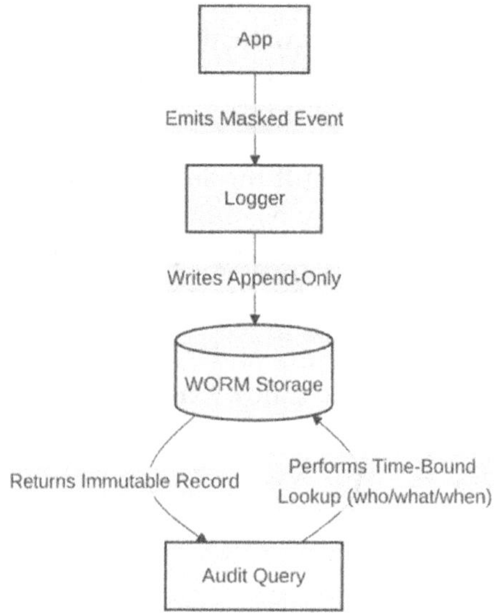

Figure 6.2*: Audit trail end-to-end (minimal)*

Key considerations

To build a trustworthy and efficient audit system, several core principles must be addressed. These considerations form the foundation for turning raw activity logs into durable, tamper-proof evidence:

- **Accountability through immutability:** The primary goal is to create an unbroken chain of evidence that ensures accountability and data integrity. By implementing WORM retention policies, organizations can guarantee that once logs are written, they cannot be altered or deleted. This transforms the audit trail from a simple record into a legally defensible and forensically sound narrative of events, allowing organizations to accelerate their audit cycles while ensuring that the underlying evidence remains tamper-proof.

- **Evidence over activity:** Standard logging captures activity, but a robust audit strategy transforms that activity into durable evidence. This involves not just collecting logs but also structuring them for long-term retention, analysis, and presentation. The objective is to prepare evidence, such as queries, data exports, and configuration changes, in a way that makes audits a matter of minutes, not weeks, by providing a centralized audit trail URL that points directly to all required documentation.

- **Comprehensive event capture:** In complex multi-cloud environments, effective auditing requires a holistic view that consolidates logs from numerous sources. It is crucial to capture a wide array of events, including user access, configuration changes, data plane operations, and administrative actions. Furthermore, supplementing automated logs with human-readable decision logs provides critical context, creating a complete narrative as outlined in the evidence matrix shown in *Table 6.1*.

Therefore, a successful audit strategy relies on making logs immutable with WORM policies, treating logs as durable evidence, and capturing a comprehensive range of events from all systems.

Process or steps

Creating a consolidated and immutable evidence package involves three key steps:

1. **Centralize artifacts:** The initial cause of audit friction is disparate evidence scattered across multiple systems. The corrective action is to consolidate all relevant artifacts, including decision logs, test results, deployment manifests, and security scan reports, into a single, designated location in cloud storage. The result is a unified source of truth that simplifies access for auditors and incident responders.
2. **Apply immutability:** To ensure the integrity of the consolidated evidence, the storage location must be made tamper-proof. The action is to apply a WORM policy, such as AWS S3 Object Lock or a Google Cloud Storage retention policy, to the bucket or container. The result is that all uploaded artifacts are protected from modification or deletion for a defined retention period, creating a durable and trustworthy record.
3. **Generate manifest and proof:** Auditors require a clear entry point and a way to verify the evidence's integrity. The action is to generate a manifest file (e.g., `index.md`) that serves as the entry point, summarizing the change and linking to all artifacts, and a WORM proof file that programmatically details the active WORM configuration. The result is a single, shareable audit trail URL that provides a complete, verifiable, and easily navigable audit package, allowing teams to move quickly through development cycles and still produce a complete, immutable audit package.

Therefore, the process involves centralizing all audit artifacts, making them immutable with a WORM policy, and creating a manifest file that serves as the single entry point for auditors.

Concrete example

A practical application of this process can be seen when a team at Fusion-Mart deploys a critical update to a production application. Following a change request, the team gathers all associated evidence into a local directory structure.

This structure includes a decision log file explaining the rationale for the update, a test results directory containing performance and security reports, a deployment manifest used for the release, and a metrics summary file outlining expected impacts. Once all files are collected, the team uploads the entire directory to a pre-configured, WORM-enabled S3 bucket. A script then generates a WORM proof file by querying the S3 bucket's Object Lock status and retention period.

Finally, a manifest file is created with a summary of the change and relative links to each piece of evidence. The access-controlled, shareable URL for this manifest file becomes the final audit trail URL submitted to the audit team. This allows the team to confidently push the deployment without creating a future compliance burden.

Therefore, a team bundles all evidence for a software update, decision logs, test results, and manifests into a directory, uploads it to a WORM-protected S3 bucket, and generates a manifest file whose URL becomes the single, immutable link for auditors.

Cautions and limits

While implementing an immutable audit trail is powerful, organizations must be aware of potential challenges and limitations to ensure its effectiveness. These are not blockers but manageable risks that require careful planning.

One significant risk is the misconfiguration of WORM policies. Setting a retention period that is too short may violate compliance requirements, while accidentally using S3 Object Lock's GOVERNANCE mode (which allows authorized users to override retention and delete objects) instead of COMPLIANCE mode (which prevents all deletions until retention expires) could allow for deletions. To mitigate this, teams should use **infrastructure as code (IaC)** to define and deploy storage bucket policies. This enables peer review and automated validation. Another concern is the cost associated with storing vast quantities of log data over long periods. Teams can leverage cost-effective tiered storage for long-term retention, moving older, less frequently accessed logs to archive tiers without losing the ability to retrieve them for forensic analysis. Finally, the system's integrity depends on the quality of the evidence collected; an automated pipeline job that validates WORM policy settings before archiving ensures the process remains robust. The differences in WORM implementations across clouds also require careful consideration.

Therefore, key risks include misconfigured WORM policies and high storage costs, which can be mitigated with IaC, peer reviews, and tiered storage strategies.

Glossary

The following are the key terms that have been discussed in this section:

- **WORM:** A data storage model that allows data to be written a single time and prevents any modification or deletion for a specified retention period, ensuring data immutability.
- **Object lock:** A feature available in cloud storage services like Amazon S3 that provides WORM capabilities. It helps prevent objects from being deleted or overwritten for a fixed amount of time or indefinitely.
- **Audit trail:** A centralized, immutable repository, typically identified by a single URL,

that consolidates all artifacts (e.g., logs, test results, decision records) required to audit a specific change.

- **Log-archive pattern:** An architectural pattern for enhancing log security and integrity. It involves sending all critical logs from operational accounts to a separate, dedicated, and highly restricted account with immutable storage policies.
- **Key management service (KMS):** A cloud service that allows users to create, manage, and control cryptographic keys across their applications and services. It is used to encrypt data at rest, including audit logs.

Expert notes

For experienced practitioners, optimizing the audit process involves deeper architectural and automation strategies. The log-archive pattern is a critical best practice; by isolating all critical logs like AWS CloudTrail and S3 access logs in a separate, hardened AWS account, the audit trail is protected even if a primary workload account is compromised. Within this archive, using S3 Object Lock in COMPLIANCE mode provides the strongest guarantee of immutability, as it prevents deletion by any user, including the root account. Automating evidence collection streamlines the developer workflow even as it strengthens the integrity of the audit trail. For instance, the WORM proof file should not be created manually. Instead, a CI/CD pipeline step should execute a script that auto-generates the WORM proof file by querying the cloud provider API for the live bucket policy, ensuring the proof is always accurate and verifiable. This approach also aligns with data vault principles by creating a clear, auditable data lineage for every change, where the audit trail acts as a satellite providing rich context.

Major cloud providers (AWS S3, Azure Blob Storage, Google Cloud Storage) offer WORM-capable storage with retention policies. These services support object-level or container-level immutability, typically with options for governance mode (privileged override allowed) or compliance mode (irreversible).

WORM storage typically offers two modes: Governance mode allows privileged users to bypass retention policies for operational flexibility, while compliance mode enforces irreversible retention with no bypass capability, suitable for strict regulatory requirements.

Therefore, experts should implement the log-archive pattern with a separate, hardened account, use COMPLIANCE mode for Object Lock, and automate the generation of the WORM proof file via API calls in the CI/CD pipeline.

Related topics

This section focuses on establishing a durable, immutable audit trail for long-term retention and efficient auditing. For related topics addressed elsewhere in this book:

- CI/CD pipeline integration, see *Chapter 8, DevOps and CI/CD for Data Engineering*
- Cloud migration strategies, see *Chapter 9, Cloud Migration and Coexistence Strategies*

Audit trail best practices for data governance

To pass stringent evidence and decision gates and maintain robust data governance, organizations must create a comprehensive and immutable trail of evidence for every change. The audit trail URL serves as this single source of truth, consolidating all decision logs, test results, and deployment artifacts into one auditable location. This approach enables rapid development cycles while still providing the rigorous, unchangeable proof required to pass stringent audits. By centralizing evidence in a tamper-proof repository, teams can demonstrate compliance efficiently, reduce audit friction, and build trust with regulators. The assurance of this process is anchored by a verifiable link to the CI/CD pipeline logs that confirm the successful aggregation and storage of all required artifacts.

Key considerations

To successfully implement an audit trail, teams must focus on three primary considerations:

- **Centralization for audit efficiency:** Consolidating all evidence into a single, stable URL is paramount. Auditors should not have to hunt for disparate logs, test reports, or approval records. A unified audit trail streamlines the review process, reduces ambiguity, and accelerates audit completion, even as it simplifies the developer's workflow. A standardized evidence manifest file within the hub acts as a table of contents, confirming all required artifacts are present.

- **Immutability for evidence integrity:** The credibility of audit evidence hinges on its integrity. Implementing a WORM policy on the underlying cloud storage is non-negotiable. This ensures that once an artifact is written, it cannot be altered or deleted for a defined retention period, providing a verifiable guarantee against tampering. This allows teams to confidently present evidence without needing to defend its authenticity. The state transition to immutability is a one-way process.

- **Auditable data lineage:** Understanding the origin and transformation of data is crucial for compliance. Applying principles from data vault, such as using hash keys and separating structural components, creates a clear and auditable data lineage. This structure ensures that every piece of data can be traced back to its source, providing a transparent history of its journey through the system and still allowing for agile data model evolution. An audit trail link showing the data's path from source to target provides concrete proof of this lineage. Beyond audit trails, effective data governance requires operational controls for data protection. RBAC ensures users access only the data necessary for their role. Data masking and column-level security protect sensitive fields (e.g., **personally identifiable information** (**PII**), financial data) from unauthorized viewing. For GDPR compliance, organizations must implement processes for right-to-be-forgotten requests, enabling the deletion or anonymization of personal data upon request. These controls complement the audit trail by ensuring that the data being governed is itself properly protected.

Therefore, a successful audit trail relies on consolidating all artifacts into a single URL, ensuring their integrity with WORM storage, and maintaining clear data lineage.

Process or steps

Creating a compliant audit trail for each change involves a systematic, automated process:

- **Aggregate all relevant artifacts:** The process begins when a change, such as a code commit or configuration update, triggers a CI/CD pipeline. The pipeline is responsible for gathering all required evidence, including source code links, static analysis reports, unit and integration test results, peer review approvals, and deployment logs. This step ensures that all activities related to the change are captured without manual intervention.
- **Commit to immutable storage:** The aggregated artifacts are then packaged and written to a designated cloud storage bucket. This bucket must be configured with a minimal WORM policy. This action creates a permanent, tamper-proof record, guaranteeing the evidence's authenticity for its required retention period, which can be defined according to a matrix like the one in *Table 6.1*.
- **Generate and distribute the audit trail URL:** Once the artifacts are securely stored, the pipeline generates a persistent URL pointing to the location of the evidence package. This URL is the final product, the audit trail. It is then automatically published to relevant systems, such as by attaching it to the corresponding Jira ticket or deployment record, providing a direct and immutable link from the change request to its complete audit trail, visualized in *Figure 6.2*.

Therefore, the process automates the aggregation of change artifacts, secures them in WORM storage, and generates a single, persistent URL for auditors.

Concrete example

A development team at Fusion-Mart needs to deploy a patch to fix a bug in their inventory management system. The change triggers their automated CI/CD pipeline, which executes a series of predefined steps to build and validate the software while simultaneously creating the audit trail. The pipeline first runs the unit tests and a security scan, saving the output logs. It then packages the application for deployment. Before deploying, it archives the test logs, scan results, the specific code commit hash, and the deployment configuration into a compressed file. This archive is uploaded to a cloud storage bucket with a WORM retention policy, ensuring it cannot be modified. The pipeline's final step is to post a comment on the associated Jira ticket containing the stable URL to this immutable archive. When an auditor later reviews the change, they can follow the audit trail URL directly from the ticket and find all verifiable proof in one place, enabling rapid validation without disrupting the development team. The assurance artifact is the immutable CI/CD job log itself, which contains the audit trail URL and confirms the successful execution of each step.

Therefore, a CI/CD pipeline can automatically compile test results and deployment logs, save them to WORM storage, and link the resulting audit trail URL to the original change ticket for seamless auditing.

Cautions and limits

While an audit trail provides a powerful mechanism for auditability, teams must be aware of potential challenges. A primary concern is the cost and complexity of long-term immutable storage; retaining vast amounts of data indefinitely can become expensive. Organizations must define a clear data retention policy to balance compliance needs with budget constraints. Another limitation is the risk of incomplete evidence; if the automated process fails to capture a critical artifact, the audit trail's value is diminished. This requires robust pipeline monitoring and validation checks to ensure all required evidence is collected for every change. Teams can move forward with implementation while mitigating these risks by having a feature flag to disable evidence collection for low-risk, non-production environments. Finally, WORM policies are rigid by design; once data is locked, it cannot be easily removed, even if it was stored by mistake, making initial setup and testing critical.

Therefore, be mindful of storage costs by setting clear retention policies and ensuring evidence completeness through robust automation, as WORM storage is unforgiving.

Glossary

The following are the key terms that have been discussed in this section:

- **Audit checkpoint:** A formal review point in a project or development lifecycle where compliance with a set of standards or regulations is verified before work can proceed.
- **Compliance mode:** A WORM retention setting that prevents any modification or deletion of stored objects, including by privileged users, ensuring the strongest level of evidence immutability for regulatory compliance.
- **Governance mode:** A WORM retention setting that allows privileged users with special permissions to modify retention settings or delete objects, providing flexibility while maintaining audit controls through authorization logs.

Expert notes

For advanced practitioners, the choice of WORM implementation carries significant nuance. Cloud providers offer different retention modes, such as the Governance and Compliance modes in AWS S3 Object Lock, which have different permissions. Compliance mode is the most stringent, preventing even the root account from altering the retention policy, and is often necessary for meeting strict regulatory requirements like SEC rule 17a-4. Furthermore, integrating the audit trail with data vault principles can create a powerful combination for auditable data lineage. By using satellite tables to track the load date and record the source of evidence artifacts, you can build a complete, verifiable history of how and when evidence was collected. This allows for rapid progress on data integration and still provides a deep, auditable history. For forensic analysis, a link to a legal hold policy in the cloud provider's console serves as a concrete assurance artifact, demonstrating the capability to preserve evidence beyond standard retention periods if required.

Therefore, advanced users should select the specific WORM mode that matches their regulatory needs and can enhance auditability by integrating audit trail artifacts into a data vault model for deeper lineage tracking.

Related topics

This section focuses on the creation and use of an audit trail URL for passing evidence and decision gates. For related topics addressed elsewhere in this book:

- Pipeline governance and deployment, see *Chapter 8, DevOps and CI/CD for Data Engineering*
- Multi-cloud IAM strategies, see *Chapter 4, Medallion Architecture in Multi-cloud*

Delta Lake transaction logs for compliance and auditing

In modern data ecosystems, maintaining data integrity and providing a transparent audit trail are essential for regulatory compliance and operational trust. The Delta Lake transaction log serves as the cornerstone for these guarantees. It functions not just as metadata but as a complete, immutable record of every operation performed on a table. This provides a detailed history of which job or service principal changed what data and when, enabling organizations to meet stringent audit requirements while still empowering teams to innovate with their data. The ability to forensically analyze data evolution and prove compliance is a critical capability that the transaction log delivers by design, with a permanent link to the log directory serving as the core assurance artifact.

Key considerations

The Delta Lake transaction log provides a robust foundation for auditing through several inherent design features. These capabilities allow organizations to build reliable compliance and governance workflows directly on top of their lakehouse architecture:

- **Immutable transaction history:** The transaction log is an append-only, ordered series of files that records every change to a Delta table. This immutability ensures that once an operation is committed, its record cannot be altered or removed, creating a tamper-proof audit trail. This characteristic is fundamental for forensic analysis and compliance, providing an unassailable history of data modifications and forming the basis for a WORM data governance policy.

- **Historical state reconstruction:** Delta Lake leverages the transaction log to enable its time travel feature, allowing users to query the state of a table at any previous version or point in time. For auditors, this is a powerful tool for root cause analysis, as it allows them to reconstruct the exact state of the data before and after a specific event. This capability simplifies the process of identifying the source of data discrepancies and validating remediation efforts.
- **Direct log analysis:** The transaction log consists of queryable JSON files that contain detailed metadata about each operation, including user identity, operation type, and parameters. This structure permits auditors to perform flexible and granular analysis directly on the log itself, moving beyond high-level summaries to investigate specific events. Teams can develop custom queries to monitor for specific activities, such as schema changes or large-scale deletions, and still maintain a rapid development pace, with a saved query that flags unauthorized operations serving as a key control.

Therefore, the Delta Lake transaction log's core design principles of immutability, time travel, and queryable metadata provide a powerful, built-in toolkit for comprehensive data auditing.

Process or steps

Auditing a data discrepancy in a Delta Lake table involves a systematic, three-step process that leverages the transaction log to move from a high-level overview to a precise root cause:

- **Establish operational context:** Begin by using the `DESCRIBE HISTORY` command to get a chronological overview of all transactions committed to the table. This initial step helps identify the scope of the investigation by revealing who performed which operations (e.g., `WRITE`, `MERGE`, `UPDATE`) and when they occurred. This high-level view, detailed in *Listing 6.1*, allows an auditor to quickly pinpoint suspicious activities or timeframes that correlate with the reported issue.
- **Isolate the causal change:** Once a specific transaction or timeframe is identified, directly query the underlying JSON log files to inspect the detailed operational metadata. By parsing fields like `operationParameters` or `schemaString`, you can uncover the exact logic that was executed, such as the join condition in a `MERGE` statement or the new columns added during a schema evolution. This granular analysis, shown in *Listing 6.2*, is crucial for isolating the precise root cause of the data discrepancy.
- **Verify and quantify impact:** After identifying the problematic transaction, use Delta Lake's time travel capability to query the table's state immediately before and after the event. This comparative analysis, demonstrated in *Listing 6.3*, allows you to

definitively confirm the impact of the change by comparing the healthy historical data with the current, erroneous data. This step is essential for quantifying the scope of the issue and planning for accurate data correction, all while a versioned audit notebook provides a repeatable and verifiable process.

Therefore, a systematic audit follows a path from a high-level history review to exploring specific transaction metadata, concluding with time travel to verify the exact impact.

The following code listing shows the two methods for inspecting a Delta table's transaction log, the source of its audit trail:

```
-- Preferred: High-level, human-readable summary of all table transactions.
DESCRIBE HISTORY delta.`/mnt/delta/bronze/customer_orders`;

-- Alternative: Direct query of the raw JSON log files for deep inspection.
SELECT * FROM json.`/mnt/delta/bronze/customer_orders/_delta_log/*.json`;
```

Listing 6.1: *Two methods for inspecting a Delta table's transaction history*

The following code listing shows querying the Delta transaction log to audit historical schema changes:

```
SELECT
  version,
  timestamp,
  operationParameters.partitionBy AS partitioning_scheme
FROM
  json.`/mnt/delta/bronze/customer_orders/_delta_log/*.json`
WHERE
  operation = 'WRITE' AND operationParameters.mode = 'Overwrite'
```

Listing 6.2: *Querying the Delta transaction log to audit historical schema changes*

The following code listing shows reconstructing a table's historical state using Delta Lake's point-in-time query capabilities:

```
SELECT *
FROM delta.`/mnt/delta/gold/products` VERSION AS OF 5;

SELECT *
FROM gold_products TIMESTAMP AS OF '2025-09-01T12:00:00Z';
```

Listing 6.3: *Reconstructing a table's historical state*

Concrete example

Imagine an auditor at Fusion-Mart investigating a sudden drop in reported sales figures from the **`daily_sales`** table. They suspect an erroneous data update corrupted the records. The auditor follows the established process to diagnose the issue. First, they execute **`DESCRIBE HISTORY daily_sales`** to review recent transactions, as shown in *Listing 6.1*. The output reveals a **`MERGE`** operation that occurred shortly before the discrepancy was reported, executed by an automated job. To investigate further, the auditor queries the specific JSON log file for that transaction version. They discovered the **`operationParameters`** for the **`MERGE`** statement contained an incorrect join condition, causing valid sales records to be incorrectly overwritten. Finally, to confirm the root cause and plan a fix, the auditor uses time travel to query the table state from a version just before the faulty **`MERGE`** operation, as demonstrated in *Listing 6.3*. This query returns the correct sales figures, confirming the source of the error and providing the necessary data for remediation. This entire investigation can proceed quickly without disrupting ongoing data ingestion, with a data quality dashboard serving as the definitive artifact showing the table's health before and after the fix.

Therefore, by using **`DESCRIBE HISTORY`**, direct log inspection, and time travel, an auditor can efficiently trace a data discrepancy from a high-level symptom to a specific, verifiable root cause.

Cautions and limits

While the Delta Lake transaction log is a powerful audit tool, it is important to understand its operational boundaries and to ensure its effectiveness. The **`VACUUM`** command, used for data retention and cost management, permanently deletes data files that are older than a specified retention period. Aggressive **`VACUUM`** policies can limit the time travel window, potentially removing historical data needed for an audit. To mitigate this, organizations must set a retention period that aligns with their compliance requirements, ensuring data is accessible for forensic analysis even as routine cleanup occurs. Additionally, accessing the transaction logs requires appropriate permissions; securing the **`_delta_log`** directory with strict IAM policies is critical to prevent unauthorized access or tampering. A properly configured IAM policy that grants read-only access to auditors is a non-negotiable assurance artifact. These manageable challenges highlight the need for deliberate configuration rather than inherent limitations.

Therefore, effective use of the transaction log for auditing requires balancing data retention policies with storage costs and implementing strict access controls to protect log integrity.

Glossary

The following are the key terms that have been discussed in this section:

- **Delta Lake transaction log (_delta_log):** The ordered, atomic record of every transaction applied to a Delta Lake table. It is the single source of truth that defines the table's state at any point in time, enabling ACID transactions and time travel.

- **Immutability:** The property of an object or record that prevents it from being modified after it has been created. The append-only nature of the Delta transaction log ensures that historical records are tamper-proof.
- **Time travel (Delta Lake):** A feature that allows users to query a previous version of a Delta table using a version number or timestamp. It leverages the transaction log to reconstruct historical data states for auditing, rollbacks, or analysis.

Expert notes

For advanced auditing and governance, teams can move beyond manual inspection and build automated monitoring on top of the transaction log. By programmatically parsing the JSON log files in real-time, organizations can create alerts for specific events, such as unauthorized schema changes, large-scale data deletions, or modifications by specific users. This proactive approach enables rapid response to potential compliance violations or data quality issues. Furthermore, to meet long-term retention requirements for regulated industries, the transaction logs can be periodically archived to immutable cloud storage. Configuring a WORM policy on a storage bucket creates a legally defensible, long-term evidence store. This allows for efficient data management on the active table while still ensuring a complete, unalterable history is available for future audits, with a feature flag to control this archival process acting as the key operational control.

For teams using Databricks, Unity Catalog provides a comprehensive data governance layer that integrates with Delta Lake. It offers centralized access control, data lineage tracking, and audit logging across all workspaces. Unity Catalog's fine-grained permissions and data discovery capabilities complement the audit trail pattern, providing both operational governance and audit evidence in a unified platform.

Therefore, advanced users can leverage programmatic parsing of transaction logs for automated monitoring and archive them to WORM-compliant storage to create a robust, long-term audit record.

Related topics

This section focuses on using the Delta Lake transaction log as an audit trail. For related topics addressed elsewhere in this book:

- Delta Lake time travel for recovery, see *Chapter 5, Building Scalable Data Pipelines*
- Medallion architecture design, see *Chapter 4, Medallion Architecture in Multi-cloud*

Navigating BFSI, healthcare, and retail regulatory mandates

Establishing a robust and auditable change management process is critical for maintaining data integrity and meeting regulatory requirements. The primary goal is to pass the formal evidence and decision gate by creating a comprehensive and immutable audit trail URL for every significant change. This URL acts as a permanent, consolidated record containing all decision logs, test results, and other essential artifacts. Achieving this requires implementing a minimal WORM policy on cloud storage to ensure data immutability, thereby creating a tamper-proof audit trail. This approach enables rapid development cycles while ensuring that every change is fully documented and verifiable for audit purposes. A single, permanent audit trail URL serves as the concrete assurance artifact for each change.

Key considerations

To build a successful and auditable system, several core principles must be addressed from the outset. These considerations form the foundation for a compliant and efficient change management process:

- The central objective is to create a single, immutable audit trail URL for each change to pass the evidence and decision gate. This URL must point to a permanent, unchangeable collection of evidence that justifies and validates the change. This ensures that auditors have a complete and trustworthy record, streamlining the review process and minimizing compliance risks. Teams can innovate on data models and pipelines even as they build a comprehensive audit trail that satisfies stringent compliance checks.

- Achieving a comprehensive audit trail requires consolidating all relevant artifacts into one accessible location. This includes everything from initial proposals and decision logs to code reviews, automated test results, and deployment confirmations. The audit trail URL links to this complete package, providing a holistic view of the change lifecycle as detailed in the evidence matrix shown in *Table 6.1*. The audit trail URL itself, containing links to all required documentation, acts as the primary assurance artifact.

- Data immutability is the technical cornerstone of a trustworthy audit trail, as illustrated in the end-to-end audit flow in *Figure 6.2*. Implementing a WORM policy on a cloud storage service prevents the accidental or malicious alteration or deletion of evidence. This is complemented by applying data vault principles to model data lineage, ensuring that the origin and transformation of all data are clearly documented and auditable, which supports the overall goal of verifiable data integrity.

Therefore, the strategy hinges on creating a single, immutable audit trail URL for each change by consolidating all artifacts and ensuring their permanence with WORM storage.

Process or steps

Creating a compliant audit trail URL for each change involves a systematic process to gather, store, and present audit artifacts:

- **Consolidate change artifacts:** For any proposed change, gather all related evidence into a single, organized structure. This includes business requirements, decision records, pull requests, automated test reports, security scan results, and deployment logs. This structured process allows for the continuous deployment of data changes without compromising the integrity or auditability of the historical record.

- **Store evidence in WORM storage:** Upload the consolidated artifacts to a designated cloud storage location where a WORM policy is enforced. This policy ensures the evidence becomes immutable, meaning it cannot be altered or deleted for a predefined retention period. Configuration examples for creating this tamper-proof archive are available for different cloud providers. A feature flag controlling the new data pipeline, coupled with a documented rollback plan within the audit trail, provides a concrete safety mechanism.

- **Apply lineage principles and finalize URL:** Structure the data and metadata according to data vault principles to ensure clear, auditable lineage, as conceptualized in *Figure 6.1*. Once all artifacts are stored and protected by the WORM policy, generate a permanent URL to the top-level directory or manifest file. This audit trail URL becomes the official, immutable record of the change, ready for submission to the evidence and decision gate.

Therefore, the process involves consolidating all change artifacts, securing them in immutable WORM storage, and generating a permanent URL that serves as the complete audit record.

Concrete example

A data engineering team at Fusion-Mart needs to add a new **`email_opt_in`** column to the customers table to support a new marketing initiative. To comply with audit requirements, they follow the established process to create an audit trail URL. First, they gather all artifacts: the project ticket describing the business requirement, the architecture decision record approving the schema change, the pull request containing the SQL migration script, and the CI/CD pipeline logs showing that all unit and integration tests passed. The team can deploy this schema change quickly and still provide auditors with a complete, tamper-proof record of the decision-making and validation process.

Next, they place these artifacts in a dedicated prefix within an Amazon S3 bucket. This bucket was configured at creation with S3 Object Lock to enforce a WORM. This action makes the stored test logs and decision records immutable. Finally, they document the S3 URI for this prefix, which serves as the permanent audit trail URL. This URL is attached to the project ticket, providing a single, verifiable link for auditors to review the entire history and validation of the change. The specific, versioned S3 URL for the change's **`README.md`** file serves as the immutable audit trail link.

Therefore, a team adding an **`email_opt_in`** column would gather the ticket, code review, and test logs, store them in a WORM-enabled S3 bucket, and use the resulting URL as their immutable audit evidence.

Cautions and limits

Implementing an immutable audit trail requires careful planning to avoid potential pitfalls. The transition to immutability is a one-way, irreversible process, making the initial configuration critical. One primary caution is the irreversible nature of WORM policies on cloud storage; a misconfigured retention period could lead to excessive storage costs or premature data deletion if not set correctly. It is essential to define and verify these policies using IaC with peer reviews before applying them. While implementing an immutable audit trail requires careful initial setup, it ultimately simplifies audit preparation without hindering development velocity.

A second limitation is the dependency on complete evidence collection. An immutable record of incomplete information is of little use for an audit. Teams must use strict checklists, such as the one exemplified in *Table 6.1*, or automated pipeline gates to ensure all required artifacts are present before the record is sealed. Finally, the choice between different retention modes, such as governance and compliance, has significant implications for administrative control. Governance mode allows certain users to override settings, offering flexibility, while compliance mode offers the strongest guarantee of immutability, which may be required by regulation. A quarterly review process ensures that retention policies and access controls remain appropriate.

Therefore, the key challenges include the irreversibility of WORM policies, the risk of sealing incomplete evidence, and selecting the correct retention mode to balance flexibility and compliance.

Glossary

The following are the key terms that have been discussed in this section:

- **RACI matrix:** A responsibility assignment framework that defines who is responsible (executes the task), accountable (makes final decisions), consulted (provides input), and informed (receives updates) for each governance process or decision.

Expert notes

For experienced practitioners, optimizing the evidence collection and management process can yield significant efficiency gains. A primary best practice is to automate evidence gathering directly within the CI/CD pipeline. The pipeline can be scripted to collect test manifests, code coverage reports, and deployment receipts automatically, packaging them and depositing them into the WORM storage location. This reduces manual effort and ensures consistency. Advanced teams can fully automate the creation of these audit packages, allowing them to accelerate delivery while providing an even higher degree of compliance assurance.

Furthermore, aligning the collected evidence with specific regulatory controls provides a direct link between technical work and business compliance. This can be achieved by creating a mapping that connects each artifact type to the regulation it helps satisfy. It is also crucial to establish clear governance procedures for managing WORM policies, including break-glass scenarios. Using S3 Object Lock's governance mode provides a safety net, while a formal review process governed by a RACI matrix ensures that any overrides are deliberate and authorized. An automated CI/CD pipeline job that fails if any required evidence artifact is missing serves as a proactive assurance mechanism.

Map each control to specific regulatory clauses and required evidence. For example, retention policies may require bucket policy JSON and lock proof, while access reviews may require IAM diffs and ticket IDs.

Establish clear ownership for critical processes such as break-glass procedures (typically requiring CISO approval) and retention policy changes (requiring data governance responsibility and CIO approval).

Therefore, experts should automate evidence collection via CI/CD, map artifacts directly to regulatory controls, and establish clear governance for managing WORM policies and exceptions.

Related topics

This section focuses on the principles and general process for creating an immutable audit trail. For related topics addressed elsewhere in this book:

- Data vault modeling principles, see *Chapter 2, Understanding Data Mesh, Lakehouse, and Medallion*
- Cloud platform implementation, see *Chapter 9, Cloud Migration and Coexistence Strategies*

Lab: Data governance with a tiny data vault

In this lab, we will explore data vault modeling principles to detect governance gaps. You will take on the role of a governance auditor for Fusion-Mart, investigating transactions that reference policies not properly modeled in the data vault. By directly inspecting hub and satellite structures, you will gain firsthand experience with how data vault captures business keys and attributes, enabling comprehensive governance validation and gap detection.

Lab environment: All labs run in GitHub Codespaces with a dev container. Every lab produces a standardized `labs/ch06/artifacts/result.json` containing status, checks, metrics, and messages, ensuring consistent governance evidence regardless of technology.

Note: **The lab uses accept/reject for simplicity. In the book text, we use Go/Retry/Kill to describe the same decision workflow.**

Conclusion

This chapter equipped you with the essential strategies to confidently pass any evidence and decision gate. By consolidating decision logs, test results, and artifacts into a single, immutable audit trail URL and implementing WORM policies on cloud storage, you have established a robust, auditable record. Furthermore, leveraging data vault principles ensures transparent and verifiable data lineage, critical for compliance.

With a solid foundation in data governance and auditable practices now firmly established, our journey continues into the operationalization of intelligent systems. The next chapter will explore how to seamlessly deploy, manage, and continuously monitor your machine learning and AI models, ensuring their ongoing performance and compliance within your governed data ecosystem.

Fusion-Mart can now answer any auditor's question with a single URL, transforming compliance from a burden into a competitive advantage.

Chapter 7
MLOps for AI Model Deployment and Monitoring

Introduction

When your ML pipeline proposes a model update, how do you know it is safe to deploy? This chapter answers the first question: *Can we trust AI-generated proposals?* through the AI proposal gate.

Deploying and monitoring AI models reliably in production presents significant governance and operational challenges, often hindering the full potential of **machine learning operations** (**MLOps**). Common challenges include silent model drift eroding business metrics, unreproducible builds making debugging impossible, and ad-hoc deployments bypassing validation. This chapter addresses these challenges by demonstrating the AI proposal evaluation pattern from the Four-Gate Governance Model (a governance framework consisting of boundary, RB-30, evidence, and AI proposal gates, introduced in the *Reading Guide*), focusing on ML model deployments, a critical instance where AI systems (trained models) must be validated before production use. The chapter explores MLOps practices, including model training, containerization, and deployment, while the lab demonstrates how these deployment decisions can be formalized as AI-generated proposals, structured change requests created by automated systems or ML engineers, then validated by governance gates before execution, subject to automated governance evaluation. While the broader AI proposal evaluation pattern encompasses validating any AI-generated artifacts (code, configurations, schemas), this chapter specifically addresses ML model governance, showing how the same validation principles apply when the proposal is a trained model rather than generated code. Model proposals are validated against predefined KPIs (accuracy, latency, throughput), organizational policies (data privacy, fairness requirements), and operational thresholds (resource consumption, rollback capability).

Structure

This chapter covers the following key topics:

- Data engineer's role in MLOps workflow orchestration
- Effective collaboration with data scientists and ML engineers
- MLOps tools and CI/CD integration
- Integrating medallion data layers into ML pipelines
- Model drift detection and rapid rollback strategies
- Lab: AI-generated change pack evaluator

Objectives

By the end of this chapter, you will be able to containerize machine learning models using Docker to achieve reproducible deployments across development, testing, and production environments and deploy ML models as managed API endpoints that enable versioning, traffic management, and instant rollback capabilities. You will establish comprehensive monitoring for production models that tracks accuracy, data drift, and performance degradation, while applying Four-Gate checkpoints to ML model releases, ensuring trained models receive the same validation as human-authored code. Finally, you will implement the AI proposal gate to validate that AI-generated changes meet the same quality standards as human work.

Data engineer's role in MLOps workflow orchestration

A structured release process is essential for deploying ML models reliably and at scale. By applying a formal governance framework, organizations can ensure that models moving to production are robust, reproducible, and aligned with business objectives. This approach transforms model deployment from an ad-hoc, risky task into a systematic, auditable workflow that balances innovation with operational stability. It establishes clear quality checkpoints for data, code, and performance, ensuring that a model can deliver consistent value even as it progresses through complex development cycles. A version-controlled deployment manifest, such as a Kubernetes deployment specification with model image tags, resource limits, and health check endpoints, or a model registry entry containing artifact URIs, input/output schemas, and performance baselines, serves as a concrete record of the exact model and configuration released.

Key considerations

Before implementing a governed MLOps workflow, consider these foundational principles:

- **Standardization through containerization**: Packaging ML models into containers, such as Docker, is fundamental for achieving reproducibility. This practice eliminates environment-specific dependencies, ensuring the model operates consistently across development, testing, and production systems, while a signed container image provides a verifiable artifact for deployment. A standardized container ensures you can advance model development without introducing difficult-to-diagnose environmental inconsistencies.
- **Phased quality verification**: A multi-stage review process introduces critical checks at logical points in the model lifecycle, from initial design to final release. This prevents low-quality assets from moving downstream, saving significant rework and reducing production risks. You can maintain development velocity and still enforce quality standards by using an automated checklist in a pull request template to confirm that all criteria for a stage are met.
- **Models as managed endpoints**: Deploying models as **application programming interface (API)** endpoints treats them as managed software components, enabling seamless integration with other applications. This approach simplifies consumption, abstracts the underlying complexity of the model, and facilitates robust monitoring and management. Deploying a new model version via a blue-green deployment strategy allows for instant rollback, ensuring service continuity.

Therefore, effective MLOps hinges on containerization for reproducibility, phased quality checks to prevent errors, and API-based deployment for operational management.

Process or steps

Implementing a governed release process involves four distinct stages from conception to production:

1. **Initiation**: The process begins when a business need is identified. At this stage, the team defines the project's scope, success criteria, and initial data requirements in a boundary configuration file. This ensures that a project proceeds with clear goals and an understanding of data availability and still allows for agile adjustments as the project evolves.
2. **Build readiness**: Once the project is approved, data engineers and ML engineers collaborate to prepare the necessary assets for model development. This involves creating data pipelines to produce high-quality training data from the curated gold layer and setting up a reproducible development environment. A successful readiness review, confirmed by a passing CI pipeline that validates data access and schema, signals that the team can begin experimentation without foundational data issues.
3. **Pre-production verification**: After a candidate model is trained, it undergoes rigorous validation before being cleared for production. The candidate (challenger) model is compared against the current production (champion) model using metrics such

as AUC, precision, recall, and latency percentiles, a pattern known as challenger-champion comparison. This stage involves checking model performance metrics, code quality, data lineage, and the containerized model's integrity. You can automate this verification using a dedicated CI job that generates a comprehensive audit report, ensuring the model is effective and secure before it faces production traffic.

4. **Production release**: With all checks passed, the model is deployed into the production environment. The deployment process places the containerized model behind a stable API endpoint and activates a comprehensive monitoring plan. Using a feature flag to manage the release allows teams to deploy the new model confidently while retaining the ability to instantly revert to the previous version if issues arise.

Therefore, the process moves from defining scope and preparing data to validating the model and finally deploying it as a monitored API endpoint.

Concrete example

Consider Fusion-Mart's goal to deploy a new fraud detection model. They follow the four-stage process to ensure a reliable release. First, in the *initiation* stage, the product manager and lead ML engineer create a boundary configuration document defining the model's objective: to identify fraudulent transactions with greater than 95% precision. The contract specifies required data sources like **`transaction_streams`** and **`user_profiles`**. This initial step ensures project alignment without prematurely over-engineering the solution. A link to this version-controlled contract serves as the project's foundational artifact.

Next, in the *build readiness* stage, a data engineer builds a pipeline that cleans and transforms raw transaction data into a **`gold_transactions`** table suitable for training. The ML engineer then uses this data to develop and train a model. The team proceeds to the next stage only after an automated data validation job confirms the **`gold_transactions`** table meets quality standards, ensuring model development can advance and still rely on a stable data foundation.

During *pre-production verification*, the trained model is containerized. An automated CI/CD pipeline runs unit tests, evaluates the model's performance against a holdout dataset, and scans the container image for security vulnerabilities. The pipeline generates a verification summary report, providing a clear, auditable record that the model is performant and secure.

Finally, in the *production release* stage, the validated container is deployed to the production environment as a REST API. The release is managed with a feature flag, initially routing only 1% of traffic to the new model. This allows the team to monitor its real-world performance closely while minimizing risk, providing a safe path to a full rollout.

Therefore, Fusion-Mart applies the process by defining requirements in a contract, preparing validated data, running automated checks on a containerized model, and deploying it cautiously using a feature flag.

Cautions and limits

Implementing a structured MLOps process introduces manageable challenges that require careful consideration. The governance framework can add overhead, potentially slowing down initial development cycles if not properly automated. To mitigate this, teams should invest in robust CI/CD pipelines that automate checks and validations, allowing them to enforce standards without creating manual bottlenecks. A well-designed pipeline can execute these checks automatically on every code commit, providing rapid feedback. Furthermore, models in production are susceptible to performance degradation from data drift (changes in the statistical properties of input data).

This risk can be managed by establishing a comprehensive monitoring plan that tracks both model metrics and data distributions. Quality standards such as **Population Stability Index (PSI)** help quantify distribution shifts, a PSI above 0.2 typically indicates significant drift requiring investigation, even as the underlying data patterns change. Teams should also monitor for label drift (changes in the target variable distribution) and account for delayed feedback scenarios where ground truth labels arrive days or weeks after predictions, requiring proxy metrics for early warning. Automated alerts, configured through a monitoring dashboard configuration file, can trigger retraining workflows when performance drops below a predefined threshold. Finally, discrepancies between development and production environments can lead to unexpected behavior. Containerization is the primary defense against this, ensuring environmental consistency, while a version-controlled Docker Compose file provides a clear definition of the service configuration.

Therefore, teams must mitigate process overhead with automation, counter model degradation with active monitoring, and prevent environment-related issues through containerization.

Glossary

This section defines key terms used throughout the discussion of the MLOps workflow:

- **API endpoint**: An API endpoint is a specific URL where an application can access a web service. In MLOps, models are often exposed as API endpoints to make predictions.
- **Containerization**: The process of packaging an application and its dependencies into a standardized unit, called a container. This ensures the application runs consistently across different computing environments, a process you can manage while maintaining security through a signed container image.
- **CI/CD**: CI/CD is a set of practices that automate the building, testing, and deployment of software. A CI pipeline configuration file defines the automated steps for validation and release.
- **Data drift**: The phenomenon where the statistical properties of the data a model sees in production change over time compared to the data it was trained on. This can degrade model performance.

- **Feature flag**: A software development technique that allows teams to turn certain functionality on or off during runtime without deploying new code. This is useful for canary releases and A/B testing.

Expert notes

For teams looking to mature their MLOps practices, several advanced techniques can enhance the governed workflow. Adopting GitOps principles, where the Git repository is the single source of truth for both application and infrastructure configuration, provides a powerful audit trail and simplifies rollbacks. Every change to the production environment, from a model update to a resource adjustment, is managed through a pull request, which serves as a verifiable change record. This allows you to accelerate deployments and still maintain strict control over the production state. Note that while routine promotions can be fully automated, high-risk changes (new model architectures, significant threshold modifications) typically require human approval as a final gate before production deployment.

Furthermore, instead of a simple full-scale deployment, consider implementing canary releases or shadow deployments. A canary release exposes a new model version to a small subset of production traffic, allowing for real-world performance evaluation with minimal risk. A shadow deployment, conversely, runs the new model in parallel with the old one, comparing their outputs without affecting user-facing results. These strategies are often managed through a service mesh configuration file, which provides fine-grained traffic control.

Finally, integrate model explainability and fairness audits directly into the pre-production verification stage. Tools like **SHapley Additive exPlanations** (**SHAP**) or **Local Interpretable Model-agnostic Explanations** (**LIME**) can generate reports on feature importance and model behavior, ensuring that models are not only accurate but also transparent and fair. This automated analysis, captured in a fairness report, becomes a critical artifact for regulatory compliance and building trust in the system.

Therefore, advanced MLOps can be achieved by using GitOps for declarative management, canary deployments for safer rollouts, and automated explainability audits for model transparency.

Related topics

This section focuses on the governance framework for deploying and monitoring a single machine learning model. For related topics addressed elsewhere in this book:

- Data engineering pipelines with dbt, see *Chapter 5, Building Scalable Data Pipelines*
- Data lineage and audit trails, see *Chapter 6, Data Governance and Compliance*

Effective collaboration with data scientists and ML engineers

Successfully transitioning a machine learning model from experimentation to production requires a symphony of specialized expertise. The collaboration between **data engineers (DEs)**, **data scientists (DSs)**, and **machine learning engineers (MLEs)** is fundamental to this process. Without seamless integration between these roles, projects often face deployment bottlenecks, data discrepancies, and a lack of reproducibility. This section outlines the collaborative methods and communication protocols that unite these disciplines. Organizations can accelerate model delivery while maintaining high standards for governance by implementing a structured, cross-functional workflow, tracked through a version-controlled project manifest that details all dependencies and review outcomes.

Key considerations

Effective collaboration in MLOps is built on a foundation of shared resources, clear expectations, and integrated team structures. Before defining a process, teams must align on these core principles to prevent friction and silos. The following points are critical for creating a cohesive MLOps environment:

- **Unified platforms and tooling**: All teams must operate from a single source of truth for data, experiments, and models. A shared feature store allows DEs to build reliable data pipelines, DSs to use consistent features for training, and MLEs to serve the same features for inference, ensuring consistency from development to production. Teams can innovate on model architecture even as they adhere to production constraints, with adherence verified by a pre-deployment checklist signed off by all three roles.

- **Formalized cross-team contracts**: Ambiguity is the enemy of efficient handoffs. Data contracts between DEs and DSs must formally define data schemas, quality standards, and update cadences. Similarly, model contracts between DSs and MLEs must specify input/output formats, performance expectations, and resource requirements, creating a clear blueprint for production services.

- **Product-centric team structures**: Organizing teams around specific ML products or use cases, rather than functional roles, fosters shared ownership and accelerates decision-making. Embedding a DE, DS, and MLE into a single pod ensures continuous communication and builds empathy for the unique challenges each role faces, moving projects forward without creating inter-team dependencies that slow progress.

Therefore, successful MLOps collaboration hinges on establishing a unified toolchain, defining explicit data and model contracts, and structuring teams around shared product goals rather than siloed functions.

Process or steps

Implementing a collaborative workflow involves a structured sequence of checkpoints that ensures a model is viable, robust, and ready for production. This structured approach enables rapid iteration without sacrificing reproducibility, with each release candidate's integrity confirmed by a container image hash recorded in the model registry:

1. **Project scoping and feasibility**: A business need triggers the initial phase, where all three roles collaborate to define the project's scope. The DE assesses data availability and quality, the DS conducts exploratory analysis to validate model viability, and the MLE provides input on deployment constraints and infrastructure needs. The outcome is a shared understanding of the project's goals and technical requirements, captured in a boundary configuration file.
2. **Development and technical validation**: With a clear scope, the teams proceed with development. The DE builds robust data pipelines to supply high-quality training data from the gold layer. The DS develops, trains, and evaluates the model, packaging the final version and its dependencies into a reproducible container. The MLE assists in standardizing this containerization process, ensuring the model artifact is prepared for deployment.
3. **Pre-production audit and deployment**: Before release, the cross-functional team conducts a final review. The DE confirms data lineage, the DS presents model performance metrics and limitations, and the MLE demonstrates the deployment plan and monitoring strategy. Upon approval, the MLE deploys the containerized model as an API endpoint and activates the monitoring dashboards, completing the transition to production.

Therefore, a successful model release follows a three-phase process of scoping, development, and auditing, where DE, DS, and MLE collaborate at each checkpoint to ensure alignment and technical readiness.

Concrete example

A financial services company aims to build a customer churn prediction model. The project begins with a scoping meeting where the DE confirms that customer activity data is available and can be delivered with hourly freshness. The DS performs an initial analysis and confirms that the available features have predictive power. The MLE specifies that any production model must have a mean response time under 100 milliseconds. These requirements are documented in a shared project charter. In the development phase, the DE builds a pipeline that cleans and transforms raw data into a feature set. The DS uses this data to train a gradient boosting model, achieving 92% accuracy. The DS and MLE work together to package the model into a Docker container. Before deployment, the team holds a final audit. The DE presents the data lineage graph, the DS shows the model's confusion matrix and fairness metrics, and the

MLE demonstrates the automated rollback plan using a feature flag. The team successfully deployed the model in one sprint and still met all compliance requirements, with a link to the final security scan report in the deployment ticket.

Therefore, a churn model project illustrates how defining contracts upfront, collaborating on artifact creation, and conducting a joint pre-deployment audit ensures a smooth and reliable transition to production.

Cautions and limits

While a structured, collaborative MLOps workflow provides immense value, teams must be aware of potential challenges and limitations. A primary risk is over-engineering the process, where rigid procedures stifle the creativity and experimentation essential for model development. It is crucial to balance governance with agility. Another challenge is the complexity of the MLOps toolchain; adopting too many new tools at once can overwhelm teams and create integration nightmares. Organizations can adopt new MLOps tools while mitigating the risk of vendor lock-in by focusing on open standards and ensuring a clear business case for each component. Finally, cultural resistance to cross-functional work can undermine any process. Without genuine buy-in and shared ownership, teams may revert to siloed behaviors. This can be managed by implementing a feature flag that allows for immediate rollback of the new model endpoint, giving teams confidence to deploy.

Therefore, teams must balance process rigor with agility, adopt tooling thoughtfully to avoid complexity, and actively foster a collaborative culture to prevent reverting to siloed operations.

Glossary

Understanding these terms allows teams to communicate precisely, enabling faster progress without creating misunderstandings. A shared, version-controlled glossary file in the main project repository serves as the single source of truth for project terminology:

- **MLOps**: A set of practices that aims to deploy and maintain ML models in production reliably and efficiently. It combines ML, DevOps, and data engineering.
- **Data contract**: A formal agreement between data producers (like DEs) and data consumers (like DSs) that defines the schema, semantics, quality, and service-level agreements for a dataset.
- **Model contract**: An agreement between the model developer (DS) and the model operator (MLE) that specifies the model's inputs, outputs, performance expectations, and resource requirements.
- **Feature store**: A centralized repository for storing, accessing, and managing machine learning features. It ensures consistency between features used for training and serving.

- **Model registry**: A centralized system for versioning, storing, and managing trained machine learning models. It tracks model lineage, metadata, and lifecycle stage (e.g., development, staging, production).

Expert notes

For experienced practitioners, optimizing the collaborative workflow involves deeper integration and automation. Adopting GitOps principles is a powerful next step, where the entire state of the MLOps system, from infrastructure configuration to model deployment manifests, is declaratively defined in a Git repository. This approach provides a complete, auditable history of all changes. Another advanced technique is implementing a standardized monitoring framework. Advanced teams can automate the entire release cycle and still provide clear audit trails for every decision, which are captured in an immutable audit log generated by the CI/CD pipeline for each deployment. This ensures that every model in production has consistent logging, tracing, and metric collection, simplifying operational management. This automation is a key part of the continuous cycle of value creation in MLOps.

Therefore, experts can enhance MLOps collaboration by implementing GitOps for declarative system management and standardizing monitoring frameworks to create a fully automated, auditable, and observable model lifecycle.

Related topics

This section focuses on the core collaborative workflow between DEs, DSs, and MLEs. For related topics addressed elsewhere in this book:

- CI/CD pipeline integration, see *Chapter 8, DevOps and CI/CD for Data Engineering*
- Model drift detection strategies, see the *Model drift detection and rapid rollback strategies* section in this chapter

MLOps tools and CI/CD integration

Robust MLOps, the discipline of streamlining the machine learning lifecycle, fundamentally relies on the principles of CI/CD. These practices enable the automation and reliability essential for modern model delivery. By applying a structured, multi-stage quality assurance framework, organizations can manage model releases with the same rigor as traditional software. This approach allows teams to accelerate the deployment of valuable models while ensuring stability and compliance. Every model deployed through this system is linked to an immutable commit hash in the source code repository, providing a clear and auditable trail of changes from development to production.

Key considerations

Adopting a CI/CD-driven approach for MLOps requires careful planning around automation, governance, and reproducibility. These practices allow teams to innovate quickly and still maintain a high bar for quality, with every change logged in an auditable CI/CD pipeline history:

- **Automation through CI/CD**: MLOps leverages CI/CD principles to automate the entire model delivery lifecycle. This includes not just code testing but also data validation, model training, and performance evaluation. Automating these steps reduces manual errors and ensures that every model candidate is built and tested consistently, leading to more reliable and predictable releases.
- **Structured release governance**: A formalized, multi-stage release process introduces critical checkpoints to ensure quality, security, and compliance. This structure prevents untested or underperforming models from reaching production. It provides clear review stages for stakeholders to assess model performance and business impact before a full rollout, creating a transparent and accountable workflow.
- **Reproducibility with containerization**: Containerization, using tools like Docker, is essential for creating reproducible machine learning deployments. A container packages the model, its dependencies, and the necessary runtime environment into a single, portable artifact. This ensures that the model behaves identically across all environments, from local development to production, eliminating environment-specific issues.

Therefore, effective MLOps relies on CI/CD for automation, a structured process for governance, and containerization for reproducible deployments.

Process or steps

Implementing a governed machine learning release workflow involves four distinct quality checkpoints. This structured process enables continuous delivery of new models without compromising production stability, as each promotion is governed by a signed-off digital checklist in the release management system:

1. **Initial validation**: A developer's code commit to the source repository automatically triggers the first stage. The CI system executes a series of automated actions, including static code analysis, unit tests, and data validation checks. If all checks pass, the system proceeds to train the model and package it into a versioned container, creating a testable model candidate.
2. **Pre-release review**: The containerized model candidate is automatically deployed to a dedicated staging environment that mirrors production. Here, it undergoes rigorous integration testing and performance evaluation against a predefined baseline or the

incumbent production model. The results, including accuracy, latency, and resource consumption metrics, are published for stakeholder review, leading to a decision on its readiness for a final audit.

3. **Final audit**: Following a successful pre-release review, the model candidate enters a final audit stage. This step involves automated security scans of the container, checks against organizational compliance rules, and a final review of all accumulated test results and documentation. This comprehensive verification ensures the model is not only effective but also secure and compliant before it is approved for production deployment.

4. **Production promotion**: Once the final audit is passed, the model is cleared for promotion to the production environment. The deployment is typically executed using a controlled rollout strategy, such as a canary release or blue-green deployment, to minimize risk. The new model version begins serving a small fraction of live traffic, and its real-world performance is closely monitored before traffic is gradually increased to 100%.

Therefore, the release process advances a model through automated validation, staging environment review, a final compliance audit, and a controlled production promotion.

Concrete example

Consider a financial technology company updating its machine learning model for real-time fraud detection. The team can rapidly iterate on the model even as they adhere to strict financial regulations, with a feature flag allowing for immediate rollback to the previous model version if any issues arise.

The process begins when a data scientist commits updated Python code, such as gold layer data loading scripts, to the project's Git repository. This action triggers the *initial validation* pipeline, which runs unit tests and builds a new Docker container with the trained model. Upon success, the pipeline automatically initiates the *pre-release review* by deploying the container to a staging environment. There, the new model is benchmarked against a dataset of historical transactions to ensure its fraud detection accuracy and latency meet established thresholds. The pipeline generates a performance report, which is attached to a pull request for review. A product manager examines this report, which serves as the auditable approval record, and approves the model for the *final audit*. This stage runs automated security scans on the container and verifies that all documentation is complete. Finally, with all checks passed, the model is ready for *production promotion*. The operations team uses a canary release strategy, initially routing only 5% of transaction traffic to the new model. After monitoring its performance and confirming no adverse effects, they incrementally increase the traffic over several hours until the new model handles 100% of requests.

Therefore, a fraud detection model is updated by automatically building a container, testing it in a staging environment, auditing the results, and gradually rolling it out to production traffic.

Cautions and limits

While a structured MLOps workflow provides significant benefits, teams must be aware of potential challenges. It is crucial to manage the complexity of these pipelines while ensuring they do not become a bottleneck, which can be achieved by using a dashboard that provides a single view of all monitoring alerts and pipeline statuses.

One primary challenge is the initial setup complexity. Designing and implementing a fully automated, multi-stage CI/CD pipeline requires specialized skills and a significant upfront investment in time and resources. To mitigate this, teams should start with a simple, minimally viable pipeline and incrementally add automation and governance checks over time. Another consideration is the resource overhead associated with maintaining multiple environments, such as staging and production. Using IaC and cloud-based services can help manage and optimize these infrastructure costs effectively. Finally, even a perfectly deployed model is susceptible to performance degradation over time, a phenomenon known as model drift. A robust monitoring plan is not optional; it is a critical prerequisite for tracking **key performance indicators** (**KPIs**), detecting drift, and triggering alerts for model retraining or replacement.

Therefore, teams should be prepared for initial setup complexity, infrastructure overhead, and the ongoing need for robust monitoring to detect model drift.

Glossary

This section defines key terms used throughout the discussion of the MLOps workflow:

- **Model drift**: The degradation of a machine learning model's predictive performance over time due to changes in the underlying data and relationships between variables.

Expert notes

For experienced practitioners, optimizing the MLOps workflow can yield further improvements in efficiency and reliability. Advanced teams can achieve near-instantaneous deployments and still ensure reproducibility by defining the entire environment declaratively in version-controlled configuration files.

Adopting a GitOps approach is a powerful best practice. In this model, the Git repository serves as the single source of truth for both the application code and the infrastructure definitions. All changes to the production environment are managed through pull requests, providing a clear, auditable, and version-controlled history. For deployment, consider advanced strategies beyond simple canary releases. Shadow deployments, for instance, allow you to run a new model version alongside the current one, feeding it production traffic without its predictions affecting users. This enables performance comparison in a real-world scenario without risk. Finally, a comprehensive monitoring plan should extend beyond simple accuracy metrics. It is vital to monitor the statistical properties of input data to detect drift, track prediction latency to ensure **service level objectives** (**SLOs**) are met, and observe system resource usage to prevent outages. This continuous cycle of value creation is central to MLOps.

Therefore, experts can enhance MLOps workflows by using GitOps for declarative management, shadow deployments for risk-free testing, and comprehensive monitoring that includes data distributions and system latency.

Related topics

This section focuses on applying a structured governance model to a standard MLOps workflow for model training and deployment. For related topics addressed elsewhere in this book:

- CI/CD pipeline fundamentals, see *Chapter 8, DevOps and CI/CD for Data Engineering*
- Data transformation patterns, see *Chapter 5, Building Scalable Data Pipelines*

Integrating medallion data layers into ML pipelines

High-quality data is the foundation of any reliable AI model. This section explains the critical role of the medallion architecture's gold layer in modern MLOps pipelines. By using curated, validated, and business-ready datasets for model training, organizations can significantly improve model accuracy, ensure reproducibility, and build trustworthy AI systems. This approach allows teams to accelerate model development while ensuring the final product is built on a foundation of verified, high-integrity data. A data quality report is automatically generated and attached to each model version, providing a clear audit trail of the training dataset's integrity.

Key considerations

Before integrating gold layer data into training pipelines, it is essential to understand its core attributes and benefits. The following points highlight the primary considerations for leveraging this high-quality data source:

- **Data quality and trust**: Gold layer data represents the pinnacle of data refinement. It is not merely clean; it is aggregated, feature-rich, and validated against business rules, providing a single source of truth for modeling. This curation process ensures that the data is reliable and directly aligned with the analytical questions the AI model is designed to answer.
- **Impact on model performance**: The quality of training data directly dictates model performance. By using a consistent and validated gold dataset, teams can minimize noise, reduce latent bias, and prevent data-related errors that often degrade predictive

accuracy. This clean foundation allows data scientists to focus on model architecture and tuning, knowing the underlying data is sound.

- **Foundation for governed MLOps**: Reproducibility is a cornerstone of effective MLOps. Gold layer data is crucial for achieving this, as it provides a stable, versioned asset for training, enabling consistent model builds over time. This supports a governed release process, ensuring that models deployed to production meet stringent quality standards.

Teams can confidently automate training pipelines with gold data, even as new data arrives, without compromising the integrity of the model builds. Each gold dataset is versioned, allowing for immediate rollback to a previous, stable version if an issue is detected. Therefore, leveraging gold layer data provides a trusted, high-quality foundation that enhances model performance, ensures reproducibility, and supports governed MLOps practices.

Process or steps

Integrating gold layer data into a model training pipeline follows a structured process to ensure quality and reproducibility:

1. **Source identification**: The process begins with identifying the correct gold layer dataset for the modeling task. This involves selecting a business-level table, such as `customer_360_features`, that has already been validated, aggregated, and prepared specifically for analytical consumption. The cause is the need for a reliable dataset, and this action selects a pre-approved, high-quality source.

2. **Data ingestion**: The selected gold data is loaded into the model training environment. This action provides a clean, consistent, and immutable foundation for the experiment, eliminating the data preparation variability that can plague model development. Access to gold datasets is controlled via an **identity and access management** (**IAM**) policy, ensuring only authorized training jobs can use the curated data.

3. **Model training and deployment**: The model is trained using the ingested gold data, and its performance is evaluated. The result is a more accurate and robust model that is a candidate for deployment. This is because the training data is trusted and versioned, the resulting model is reproducible, a critical requirement for containerization and reliable API endpoint deployment.

This structured process enables rapid model iteration while still maintaining strict data quality standards for every training run. The process involves identifying and ingesting a validated gold dataset to train a model, resulting in a reproducible and high-performance artifact ready for deployment.

Concrete example

Consider a team tasked with developing a customer churn prediction model. An initial version trained on less-refined silver layer data showed inconsistent performance due to data quality issues. To resolve this, the team transitions to using a curated gold layer table named **`customer_churn_training_set`**. This table contains validated features and a consistent schema, providing a stable foundation for training. The team can experiment with new model architectures and still ensure that performance improvements are due to algorithmic changes, not random fluctuations in data quality.

High-quality gold layer data is loaded directly into the training script. By pointing the training job to this trusted dataset, the resulting model's accuracy improves significantly, and its predictions become more reliable. A feature flag is used to route a small percentage of production traffic to the new model, allowing for safe, real-world validation before a full rollout. This disciplined approach, founded on high-quality gold data, ensures the deployed model is both performant and trustworthy, directly supporting the goal of containerizing and deploying reliable models.

Therefore, switching from raw data to a curated gold dataset for a churn model, by using curated gold layer data, directly improves model accuracy and ensures the build is reproducible.

Cautions and limits

While gold layer data provides immense value, it is important to recognize its limitations and prerequisites. Organizations can build a robust gold layer without stalling innovation by prioritizing datasets based on business impact. The following challenges should be managed proactively:

- **Cost and effort**: Creating and maintaining a gold layer requires a significant upfront investment in data engineering, infrastructure, and domain expertise. The process of transforming bronze and silver data into business-ready gold assets is resource-intensive. To mitigate this, organizations should start by developing gold tables for their most critical, high-impact use cases first.
- **Data latency**: The extensive validation and transformation steps involved in creating gold data can introduce latency. This means it may not be suitable for use cases requiring real-time model training on the very latest data. For such scenarios, using silver layer data might be a necessary compromise, with periodic retraining performed on gold data.
- **Potential for embedded bias**: Curation and validation do not automatically eliminate systemic bias present in the source data. If historical data reflects biased processes, the gold layer can inherit and even amplify that bias. Mitigation requires adding explicit fairness and bias checks to the data validation pipeline before data is promoted to gold. A data stewardship council must formally approve the schema and quality rules for any new gold table, providing a human-in-the-loop check.

Therefore, the primary limitations of using gold data are the high cost of creation, potential data latency, and the risk of perpetuating hidden bias if not explicitly checked.

Glossary

This section defines key terms used throughout the discussion of the MLOps workflow:

- **Medallion architecture**: A data organization pattern that structures data into bronze (raw), silver (cleaned), and gold (aggregated, business-ready) layers to progressively improve quality.
- **Gold layer**: The final, most refined layer in the medallion architecture. It contains highly curated, validated, and aggregated data ready for analytics and machine learning.
- **Reproducibility**: The ability to recreate a machine learning model and its results given the same code, data, and environment. It is a cornerstone of reliable MLOps.

Expert notes

For teams looking to maximize the value of their gold layer, several advanced practices are recommended. Advanced teams can implement fully automated, end-to-end data lineage tracking and still allow for agile updates to gold layer schemas. These techniques enhance governance, reproducibility, and pipeline stability:

- **Implement data versioning**: Use technologies like Delta Lake or other table formats that support time travel. This allows you to version your gold datasets, enabling you to pin a model training run to a specific version of the data. This practice is fundamental for achieving true reproducibility and for debugging model performance changes over time.
- **Enforce strict schema management**: Gold datasets should have strictly enforced schemas. Any changes to the schema, such as adding or removing a feature column, should be managed through a controlled process. Schema enforcement prevents unexpected data drift from breaking downstream model training and inference pipelines.
- **Maintain comprehensive data lineage**: For auditability and debugging, it is crucial to maintain clear data lineage that traces every column in a gold table back to its source in the silver and bronze layers. This provides transparency into how features were derived and helps diagnose data quality issues at their root. A Git-based workflow requires peer review and approval for any changes to the data transformation code that generates gold tables.

Therefore, advanced MLOps practices for gold data include versioning datasets for reproducibility, enforcing strict schemas to prevent pipeline failures, and maintaining data lineage for auditability.

Related topics

This section focuses specifically on the role and integration of gold layer data in model training pipelines. For related topics addressed elsewhere in this book:

- Medallion architecture and layer design, see *Chapter 4, Medallion Architecture in Multi-cloud*
- ETL / ELT engineering patterns, see *Chapter 5, Building Scalable Data Pipelines*
- Data quality contracts, see *Chapter 3, Data Integration Strategy, Business Impact, and ROI*

Model drift detection and rapid rollback strategies

In the lifecycle of an ML model, deployment is not the final step but the beginning of a critical operational phase. Once in production, a model's performance can degrade over time due to shifts in the data environment, a phenomenon known as drift. This degradation can silently erode business value and user trust. Therefore, establishing robust strategies for detecting model performance degradation and executing rapid rollbacks is essential for any mature MLOps practice. These mechanisms allow teams to innovate and deploy new models while ensuring that a safety net is in place, with a versioned model in a registry serving as a concrete assurance artifact for quick restoration.

Key considerations

Before implementing a detection and rollback system, it is crucial to understand the foundational principles that ensure its effectiveness. These considerations help frame the technical solution within the broader business context, allowing teams to deploy updates confidently and still maintain operational stability. A key assurance artifact here is a pre-configured alert that automatically notifies the on-call engineer via a dedicated communication channel when a performance metric breaches its threshold.

- **Business impact of degradation**: Model performance is not merely a technical metric; it is directly tied to business outcomes. A decline in a model's predictive accuracy can lead to tangible losses, such as decreased revenue, poor user engagement, or increased operational risk. Therefore, monitoring efforts must connect model metrics to KPIs to quantify the business impact of any degradation.

- **Comprehensive monitoring strategy**: Effective drift detection requires a multi-faceted monitoring plan that goes beyond simple accuracy checks. This involves tracking the statistical properties of input data to detect data drift, observing the distribution of model predictions for behavioral shifts, and, when possible, evaluating performance against ground truth labels. Using high-quality training data, such as curated gold layer data, establishes the critical baseline for these comparisons.

- **Infrastructure for rapid reversion**: The ability to perform a rapid rollback depends entirely on having the right infrastructure and processes in place beforehand. This includes using containerization for reproducible deployments, maintaining a versioned model registry to track all model artifacts, and implementing automated deployment pipelines. Without this preparation, responding to a performance issue becomes a slow, manual, and error-prone process.

Therefore, a successful strategy hinges on linking model metrics to business KPIs, implementing a multi-layered monitoring plan, and preparing the necessary infrastructure for automated rollbacks.

Process or steps

Implementing a systematic process for managing model degradation involves a clear sequence of events from detection to resolution. The following steps describe how an automated system can identify an issue, take corrective action, and restore the system to a known good state, enabling continuous improvement without sacrificing reliability. A core assurance artifact in this process is an automated rollback script committed to the CI/CD pipeline's repository:

1. **Detect performance degradation**: The process begins when a monitoring system observes a deviation from expected behavior (the cause). This could be a significant change in the statistical distribution of an input feature, detected by a **Kolmogorov-Smirnov** (**KS**) test, or a sudden drop in a key performance metric like precision below a predefined threshold.
2. **Trigger an automated rollback**: Upon detection, the monitoring system sends an automated alert that triggers the CI/CD pipeline (the action). The pipeline identifies the last known stable model version from the model registry and initiates a new deployment process to replace the degraded model with this stable version.
3. **Restore service and verify**: The deployment system, often using a blue-green or canary strategy, redirects production traffic to the redeployed stable model (the result). This action immediately mitigates the business impact. Post-rollback, the system runs health checks to verify that performance metrics have returned to their expected baseline, confirming the restoration of service.

Therefore, the process automates the response to model degradation by detecting metric deviations, triggering a pipeline to redeploy a stable version, and verifying the restoration of service.

Concrete example

Consider an e-commerce company that uses a machine learning model to recommend products to users. The new model version, v2, is deployed using a blue-green strategy, with the previous stable version, v1, remaining active on the blue environment. This setup allows the team to push forward with innovation even as it provides a clear path for reversion. The assurance artifact is a real-time dashboard that visualizes traffic flow, showing 100% of users being directed to the green environment where v2 is running.

Initially, v2 performs well. However, after a few days, the monitoring system detects a significant drop in the click-through rate, a key business metric. Simultaneously, it flags a severe data drift in the `user_region` feature, as a new marketing campaign has driven an unexpected influx of traffic from a previously underrepresented area. An automated alert is triggered, which initiates the rollback procedure. The load balancer instantly switches all incoming traffic from the green environment (v2) back to the blue environment (v1). Within seconds, the click-through rate returns to its normal, stable baseline, preventing further revenue loss. The MLOps team can then perform a root-cause analysis on v2 offline without impacting the live production system.

Therefore, a recommendation model's declining click-through rate, caused by data drift, triggers an automated traffic switch in a blue-green deployment, instantly restoring performance by reverting to the previous model version.

Cautions and limits

While detection and rollback systems are powerful, they are not without their challenges and limitations. Teams must proactively manage these potential issues to ensure the system remains reliable and effective. It is possible to advance system capabilities while implementing safeguards, such as a documented incident response runbook that provides clear steps for post-rollback analysis and communication.

One major challenge is setting appropriate alert thresholds. If thresholds are too sensitive, they can lead to *alert fatigue*, where frequent false positives cause operators to ignore real issues. Conversely, if they are too lenient, critical performance degradation may go undetected. Another limitation is that a rollback is a temporary fix, not a permanent solution; it reverts to a previous state but does not address the root cause of the drift. Furthermore, some forms of degradation, like gradual concept drift, can be too subtle for simple statistical tests to detect immediately, requiring more sophisticated monitoring techniques. Teams must invest time in tuning their monitoring systems and establishing a formal process for investigating the underlying causes of drift after a rollback occurs.

Therefore, teams must carefully tune alert thresholds to avoid fatigue, recognize that rollbacks are a temporary mitigation rather than a root-cause solution, and be aware that subtle drift can evade simple detection methods.

Glossary

This section defines key terms used throughout the discussion of the MLOps workflow:

- **Data drift**: A change in the statistical distribution of the input features of a model. The underlying relationships the model learned may still be valid, but it is receiving data it was not trained on.

- **Concept drift**: A change in the relationship between the input features and the target variable. The statistical properties of the inputs may not have changed, but the patterns the model learned are no longer correct.
- **Blue-green deployment**: A deployment strategy where two identical production environments, *blue* and *green*, are maintained. Traffic is directed to one while the other is updated, allowing for instantaneous traffic switching and rapid rollbacks.
- **KS test**: A nonparametric statistical test used to determine if two one-dimensional probability distributions differ. It is commonly used to detect data drift in a single feature.

Expert notes

For teams seeking to build more advanced MLOps capabilities, several practices can enhance drift management beyond basic rollbacks. Implementing shadow deployments allows a new model candidate to process live production traffic in parallel with the current production model, without its predictions being served to users. This provides a safe way to evaluate the new model's performance on real-world data and detect potential issues before a full rollout. This proactive validation helps teams move faster without compromising production stability, using a feature flag to control the shadow model's activation as a simple assurance artifact.

Furthermore, the choice of drift detection algorithm should be tailored to the data type and the potential type of drift. While statistical tests like the KS test are effective for univariate drift, more advanced techniques may be needed for detecting complex, multivariate drift. Finally, the entire detection and rollback process should be integrated into a larger feedback loop. The insights gained from a drift event should inform the next iteration of model development, potentially triggering an automated retraining pipeline. This creates a self-correcting system that continuously adapts to a changing environment.

Therefore, advanced techniques include using shadow deployments for pre-release validation, selecting appropriate multivariate drift detection algorithms, and integrating drift signals into a continuous retraining and improvement loop.

The monitoring strategy should integrate with immutable audit logs and the audit trail framework established in *Chapter 6, Data Governance and Compliance*, ensuring complete traceability. Each model deployment should generate evidence that flows into the decision log, creating an unbroken audit trail from development through the evidence gate to production.

Related topics

This section focuses specifically on the detection of model performance degradation and the execution of rapid rollbacks as a mitigation strategy. For related topics addressed elsewhere in this book:

- RB-30 rollback framework, see *Chapter 5, Building Scalable Data Pipelines*
- Audit trail integration, see *Chapter 6, Data Governance and Compliance*

Lab: AI-generated change pack evaluator

In this lab, you will evaluate an AI-generated change pack that proposes model promotion, demonstrating the AI proposal gate pattern where deployment decisions require structured validation before execution. The lab provides a state snapshot describing current models and performance thresholds, and a change pack proposing to promote a candidate model. You will validate the proposal against boundary constraints, RB-30 rollback requirements, and model quality thresholds (AUC). By modifying the change pack or snapshot, you will observe how governance gates accept safe proposals and reject risky ones.

Lab environment: All labs run in GitHub Codespaces with a dev container. This lab produces `labs/ch07/artifacts/result.json` containing validation results, checks status, model metrics, and accept/reject decisions.

Note: **The lab uses accept/reject for simplicity. In the book text, we use Go/Retry/Kill to describe the same decision workflow.**

Conclusion

This chapter demonstrated how to integrate the robust Four-Gate Governance Model into a standard MLOps workflow, ensuring controlled and reliable AI model deployments. You have gained essential skills, from containerizing models for reproducible releases and defining critical governance gates (boundary, RB-30, evidence, and AI proposal) to deploying models as API endpoints with comprehensive monitoring plans. This structured approach is vital for managing the lifecycle of production-ready AI systems.

With a solid understanding of governed MLOps, you are now equipped to build and manage AI systems that are not only performant but also secure, compliant, and maintainable. The principles and practices covered here serve as a foundational blueprint for operationalizing machine learning at scale. Continue to apply these rigorous standards, adapting them to the unique demands of your projects to foster trustworthy and impactful AI solutions.

In the next chapter, you will implement promotion to production within CI/CD pipelines, where all Four-Gate evidence is evaluated for the final deployment decision. *Chapter 8, DevOps and CI/CD for Data Engineering,* demonstrates how to design pipelines that validate evidence through a strict checklist (PII=0, budget adherence, RB-30 compliance, audit URL validation) and make Go/Retry/Kill decisions. You will learn to implement pipeline governance with guards and ownership at every stage, and foster DevOps collaboration. The lab focuses on pipeline fundamentals, validating guards and owners per stage, essential for reliable production deployments.

Fusion-Mart's fraud detection model, targeting greater than 95% precision, has passed all four governance gates and is ready for production deployment.

Chapter 8
DevOps and CI/CD for Data Engineering

Introduction

Modern data engineering requires robust DevOps practices and automated CI/CD pipelines to deploy data products safely at scale. Whether you are deploying machine learning models, data transformations, or infrastructure changes, the challenge remains the same: how do you move fast without breaking production? This chapter bridges the second question, production safety, addressed in *Chapter 5, Building Scalable Data Pipelines*, and question three, audit compliance, addressed in *Chapter 6, Data Governance and Compliance*, by automating the governance checkpoints established in those chapters. It explores the operational practices, cultural patterns, and automation strategies that enable high-velocity, low-risk deployments.

Successful production deployment depends on two interconnected capabilities. First, you need foundational pipeline infrastructure where every deployment stage has guards (automated checks like tests, linting, or security scans) and clear ownership (designated teams responsible for each stage). Without this foundation, no amount of process can ensure reliability. Second, you need sophisticated quality gates that validate evidence before production promotion, checking for data quality issues, budget compliance, rollback capability (RB-30 as introduced in *Chapter 5, Building Scalable Data Pipelines*), and audit trail completeness. This chapter covers both aspects, with the hands-on lab focusing on the foundational infrastructure.

Beyond technical automation, this chapter explores the human side of DevOps: fostering a culture of continuous learning and improvement, enabling effective collaboration across distributed and offshore teams, and building shared ownership of production quality. These cultural practices determine whether your CI/CD automation accelerates delivery or merely automates chaos.

The hands-on lab demonstrates foundational pipeline governance: you will validate that a small CI/CD pipeline definition has guards and owners defined for every stage. This basic infrastructure is the prerequisite for the sophisticated deployment practices explored throughout the chapter.

Structure

This chapter discusses the following topics:

- DevOps culture and mindset for data teams
- CI/CD pipelines and automated testing for data workflows.
- Remote and offshore collaboration strategies
- Lab: DevOps and collaboration on a tiny CI/CD pipeline

Objectives

By the end of this chapter, you will be able to design CI/CD pipelines for data workloads that automate testing and deployment while maintaining safety through quality gates and implement automated quality gates that validate data quality, budget compliance, rollback capability, and audit trails before production. You will make evidence-based Go/Retry/Kill deployment decisions by evaluating whether changes pass required gates and recording decisions immutably, while you foster a continuous learning culture where teams systematically enhance products, processes, and skills through iterative feedback. Finally, you will enable effective collaboration across distributed teams by establishing shared ownership and automating governance checkpoints.

DevOps culture and mindset for data teams

Continuous learning and improvement (**CLI**) is an organizational discipline for systematically enhancing data products, internal processes, and team skills. It operates through iterative cycles driven by performance feedback, evolving business needs, and technological advancements. In a data-centric organization, this philosophy is critical for maintaining relevance and competitive advantage. The foundation for effective CLI is a DevOps culture that promotes automation, shared ownership, and rapid feedback loops. This approach enables rapid innovation while maintaining high standards of data quality, verified through a version-controlled decision log that records all deployment approvals.

Key considerations

Successful implementation of continuous learning and improvement hinges on several foundational pillars. These elements create an environment where iteration is not only possible but is the default operational mode. The key considerations are as follows:

- **Iterative enhancement as a core process**: CLI is not a one-time project but an ongoing, iterative process. It involves constantly gathering feedback on data products and pipelines to identify areas for enhancement, from optimizing query performance to improving machine learning model accuracy. This allows teams to make small, frequent improvements and still deliver significant long-term value, with each change tracked via commit history in a source control system.

- **DevOps culture as an enabler**: A DevOps culture is the engine that powers CLI in a data environment. Principles like extensive automation through CI/CD, shared responsibility between data engineers and data scientists, and blameless post-mortems create a system for rapid, safe iteration. This culture allows teams to accelerate development cycles without compromising the stability of production systems.

- **Dual organizational structure**: A specific organizational design is a prerequisite for scaling CLI, particularly within a data mesh architecture. This involves a central platform team that builds and maintains self-service tools and a series of domain-oriented data product teams that own their data end-to-end. This structure allows domain teams to move quickly without sacrificing central governance, enforced by automated policy checks in the deployment pipeline.

Therefore, a successful CLI culture depends on combining DevOps principles with a dual organizational structure of an enabling central platform and autonomous domain teams.

Process or steps

Implementing a CLI cycle involves three key steps:

1. **Establish feedback loops**: The process begins by identifying the cause for change, which is discovered through robust feedback mechanisms. This includes technical monitoring for operational issues like pipeline failures or data quality degradation, performance metrics for business impact like model prediction accuracy, and direct feedback from data consumers. A feature flag system can be used to safely expose new features to a subset of users to gather targeted feedback.

2. **Execute iterative development**: Once an opportunity for improvement is identified, the domain team takes action by developing a solution. This work is managed using DevOps practices, where code changes are committed to a version control system, triggering an automated CI/CD pipeline as illustrated in *Figure 8.1*. This pipeline automates testing, validation, and deployment, ensuring changes are both effective and safe.

3. **Validate and deploy**: The result of the development work is a new version of the data product. Before promotion to production, it must pass an automated quality check, which validates evidence against a predefined checklist in *Table 8.1*. Teams can deploy changes frequently and still ensure stability through automated rollback capabilities triggered by monitoring alerts.

Data-related edge cases can complicate validation. Non-idempotent pipelines may produce different outputs on retry, making automated comparison unreliable. Model retraining variability, where the same training data yields slightly different model weights, requires tolerance thresholds rather than exact matching. Teams should design validation checks that account for these inherent variations while still detecting genuine regressions.

Therefore, the core CLI process involves establishing a feedback loop, using CI/CD to implement changes, and validating the results against predefined quality metrics.

Figure 8.1 illustrates typical CI/CD pipeline stages from code commit to production deployment. This conceptual illustration shows the general pattern of automated stages in modern CI/CD systems. The chapter lab focuses on a foundational aspect: validating that each stage in a pipeline definition has guards (automated checks) and designated owner teams, which is the prerequisite for implementing the sophisticated quality checks described in this chapter.

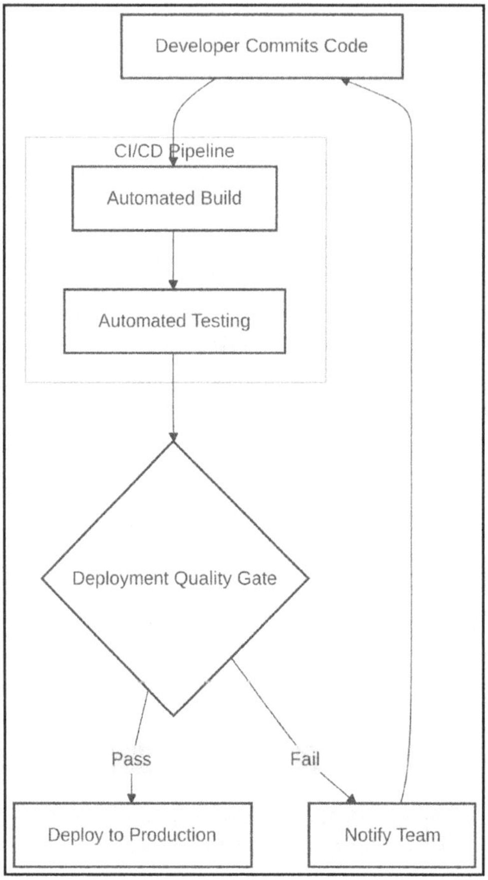

***Figure 8.1**: CI/CD pipeline from code commit to production*

The deployment quality gate, distinct from but complementary to the Four-Gate Governance Model, evaluates multiple types of evidence before allowing production deployment. *Table 8.1* shows a typical checklist used in this evaluation. Note that while this chapter explores the complete deployment workflow, including this evidence evaluation, the hands-on lab focuses on the foundational step: ensuring every pipeline stage has guards and owners defined.

Check	Status
Contract tests passed	Yes/No
SLIs within thresholds (failure_rate ≤ 1%, p95_latency ≤ 2s, retry_rate ≤ 5%)	Yes/No
Approval recorded	Yes/No

***Table 8.1**: Checklist for deployment approval*

Consider how Fusion-Mart's data engineering team deploys a pricing algorithm update. Their CI/CD pipeline automatically executes the deployment quality gate checklist: it validates that the change includes a link to the audit trail containing test results, confirms no PII exposure in the new code, and verifies that projected compute costs remain within budget. The pipeline also checks that the RB-30 rollback rehearsal was completed within the required timeframe. Only when all checks pass does the system generate a Go decision, allowing the deployment to proceed to production. This automated checkpoint ensures Fusion-Mart can ship updates rapidly while maintaining governance standards.

Concrete example

A domain team responsible for a `customer_churn_prediction` data product notices from their monitoring dashboard that the model's predictive accuracy has degraded over the last quarter. This metric serves as the feedback (the cause) that initiates a CLI cycle. The team hypothesizes that changing customer behavior has introduced model drift. To take action, a data scientist on the team develops a new version of the model with updated features and retraining on recent data. They commit their code, which triggers the CI/CD pipeline shown in *Figure 8.1*.

The automated pipeline builds the new model, runs unit and integration tests, and evaluates its performance against a holdout dataset. The results, along with other metadata, are collected as evidence. This evidence is then checked against a deployment checklist (*Table 8.1*) to ensure it meets all quality and governance standards, such as data privacy scans and budget adherence. The final decision is logged in an immutable audit trail. The team can iterate on the model quickly, even as they adhere to compliance requirements, which are confirmed by an immutable audit trail link in the deployment record.

Therefore, a domain team improves a churn model by using performance monitoring to trigger a CI/CD pipeline, which retrains and redeploys the model after passing automated quality checks.

Cautions and limits

Adopting a CLI model presents several challenges that require careful management. A primary risk is cultural resistance; teams accustomed to traditional, siloed project work may struggle with the shared ownership and rapid pace of DevOps. Another limitation is the potential for the central platform team to become a bottleneck if it cannot provide robust self-service capabilities, forcing domain teams into long wait cycles. Finally, domain teams may lack the full spectrum of skills, from data engineering to product management, required for true end-to-end ownership. Organizations can empower teams with autonomy while preventing chaos through standardized IaC templates that include security controls. Mitigating these risks requires strong leadership, investment in training, and a clear vision for the platform as an enabler, not a gatekeeper.

Therefore, key challenges to CLI adoption include cultural resistance, platform bottlenecks, and skill gaps within domain teams, which can be mitigated with training and robust self-service tooling.

Glossary

The following are the key terms that have been discussed in this section:

- **CLI**: An organizational philosophy and set of iterative practices for the ongoing enhancement of products, processes, and capabilities, driven by continuous feedback.
- **DevOps**: A set of practices that combines software **development** (**Dev**) and IT **operations** (**Ops**) to shorten the systems development lifecycle and provide continuous delivery with high software quality.
- **Data mesh**: A decentralized sociotechnical approach to share, access, and manage analytical data in complex and large-scale environments. It treats data as a product, owned by domain teams.
- **Data product**: A reusable, self-contained data asset designed to be consumed by others. It includes the data itself, the code to transform it, and the infrastructure to serve it.
- **Central platform team**: An enabling team that builds and operates a self-service data platform, providing tools, infrastructure, and governance frameworks for domain teams.
- **Domain-oriented data product team**: A cross-functional team with end-to-end ownership of the data products within a specific business domain.

Expert notes

For experienced practitioners, maximizing the value of CLI involves moving beyond basic implementation. Instead of simple monitoring (tracking known failure modes), mature teams invest in observability (the ability to ask arbitrary questions about the system without pre-defining metrics). This is crucial for debugging complex, distributed data systems. Furthermore, governance should be automated and embedded directly into the platform and CI/CD pipelines to enforce standards programmatically. This shifts governance from a manual review process to an automated quality gate, enabling speed and safety. Advanced teams can achieve high deployment velocity without compromising reliability by implementing feature flags for new data transformations, allowing for controlled rollouts and immediate rollbacks.

Therefore, advanced CLI practices include prioritizing system observability over monitoring and embedding automated governance directly into the data platform and deployment pipelines.

Related topics

This section focuses on the cultural and organizational structures that enable continuous learning and improvement in a data context. For related topics addressed elsewhere in this book:

- Data platform architecture, see *Chapter 2, Understanding Data Mesh, Lakehouse, and Medallion*
- CI/CD pipeline implementation, see the *CI/CD pipelines and automated testing* section in this chapter

CI/CD pipelines and automated testing for data workflows

In the realm of data workflows, where changes can have significant downstream impacts, the ability to release updates quickly and safely is paramount. CI/CD pipelines, coupled with robust automated testing, provide the systematic framework necessary to achieve this balance between operational agility and assurance controls. This approach transforms a potentially slow and error-prone release process into a streamlined, reliable, and auditable one. By embedding an automated checkpoint with a fixed, non-negotiable checklist into the deployment process, organizations can enforce quality and compliance standards systematically, ensuring that every change is vetted against objective criteria before it impacts production systems.

Key considerations

Successful implementation of an automated data workflow release process rests on several core principles. These considerations ensure the system is both effective and trusted by the teams who use it:

- A CI/CD pipeline provides the foundational structure for automating data workflow deployments, enabling rapid development cycles while still enforcing rigorous quality controls through integrated testing. This automated workflow, as illustrated in the deployment process in *Figure 8.1*, systematically moves code from commit to production. The primary assurance artifact in this context is a version-controlled test suite that runs automatically on every change.

- The core of the safety mechanism is an automated checkpoint governed by a fixed checklist, which ensures that every proposed change consistently meets a predefined quality and compliance standard. This removes subjectivity from the review process, guaranteeing a uniform baseline of safety for all deployments. The assurance artifact here is the published, immutable checklist itself, which provides transparency to all developers.
- This methodology codifies unwritten operational knowledge into explicit policy-as-code, a necessary step for enabling full automation and eliminating manual errors. It also fosters psychological safety by making release decisions based on objective, factual evidence rather than individual pressure or judgment. A link to the specific policy-as-code repository, included in every deployment record, serves as a concrete assurance artifact.

Therefore, a successful strategy combines a structured CI/CD pipeline with a fixed, automated checklist to enforce standards consistently, codify operational knowledge, and create a safe, objective environment for releases.

Process or steps

Executing a change through the automated deployment checkpoint involves three distinct phases:

1. **Initiate change**: A developer commits a code modification to the source control repository. This action automatically triggers the CI/CD pipeline, initiating the validation and deployment process without manual intervention.
2. **Execute checklist**: The automated system programmatically evaluates the change against a predefined checklist (*Table 8.1*). This step involves running a series of automated verifications: confirming that no **personally identifiable information** (**PII**) is inadvertently exposed, ensuring that projected compute and storage costs remain within budget, validating that a required prior risk assessment has been completed, and checking for the presence of a complete audit trail **uniform resource locator** (**URL**).
3. **Generate audit record**: Upon completing the checklist, the system makes a definitive Go/Retry/Kill decision and writes it to an immutable decision log file. This log serves as the permanent, tamper-proof audit trail. It records the final decision, the identity of the approver (human or automated), a timestamp, and a direct link to the supporting evidence, enabling teams to move forward with deployments and still maintain a complete record for compliance. The cryptographically signed decision log is the key assurance artifact of this step.

Therefore, the process is a clear sequence where a code commit triggers an automated evaluation against a fixed checklist, resulting in a final, auditable decision recorded in a log file.

Concrete example

A data science team needs to deploy an updated customer churn prediction model. The developer commits the new model code, which automatically starts the CI/CD pipeline shown in *Figure 8.1*. The deployment checkpoint then executes its checklist, detailed in *Table 8.1*:

- First, an automated scanner analyzes the model's output logs and confirms no PII is present.
- Second, a cost estimation tool analyzes the model's resource requirements and verifies that they are within the project's allocated budget.
- Third, the system checks the associated project management ticket and confirms that the mandatory peer review for architectural changes has been completed and approved.
- Finally, it validates that the commit includes a valid URL to the team's audit trail, where performance metrics and validation reports are stored.

Since all checks pass, the system generates a decision log file with a Go status. This action allows the pipeline to proceed with the production deployment, delivering business value quickly without compromising governance. The direct link to the audit trail within the final deployment ticket serves as the tangible assurance artifact.

Therefore, when a team deploys a new model, the automated checkpoint verifies it against PII, budget, review, and documentation requirements before generating a Go decision and deploying it to production.

Cautions and limits

While an automated deployment checkpoint provides significant benefits, teams must be aware of its inherent limitations and manage them proactively. The fixed nature of the checklist, for instance, may not account for unique edge cases, potentially blocking valid changes or missing novel risks. To mitigate this, organizations should establish a formal process for periodically reviewing and updating the checklist based on operational experience and incident post-mortems. Another risk is the potential for teams to treat the checklist as a bureaucratic hurdle to be cleared, for example, by linking to an empty audit trail. This can be managed by designing more robust checks that validate the content of evidence, not just its existence. Teams can deliver features rapidly, even as they must contend with these constraints. A mandatory quarterly review of the checklist's effectiveness, with outcomes tracked in a central governance log, provides a crucial assurance artifact against process stagnation.

Therefore, be mindful of checklist rigidity and the risk of superficial compliance; mitigate these by establishing a formal review process and designing robust, content-aware checks.

Glossary

The following are the key terms that have been discussed in this section:

- **CI/CD**: The practice of frequently integrating code changes from multiple developers into a shared repository and then automatically building, testing, and preparing those changes for release to a production environment.
- **Audit trail**: A centralized, accessible repository for storing all artifacts related to a change, such as test results, security scan reports, design documents, and approvals, to create a comprehensive audit trail.
- **PII**: Any data that could be used to identify a specific individual, such as a name, address, or government ID number. Its handling is often subject to strict regulatory compliance.
- **Policy-as-code**: The practice of managing and defining policies (such as compliance rules, security standards, or operational checklists) using code, enabling automation, versioning, and systematic enforcement.

Expert notes

For experienced practitioners, the concept of a fixed checklist can be extended to build more sophisticated and resilient systems. While a single, fixed checklist is a powerful starting point, mature organizations can implement context-aware checkpoints. These systems apply different, dynamically selected checklists based on the risk profile of a proposed change; for example, a simple documentation update might face a lenient review, whereas a change to a core financial dataset would trigger a much more stringent set of verifications. To ensure the integrity of the entire process, both the decision log and the contents of the audit trail should be stored in immutable or **write-once-read-many (WORM)** storage. This prevents tampering and guarantees a high-fidelity audit trail. Advanced teams can accelerate low-risk changes with these dynamic systems and still provide deeper scrutiny for high-risk modifications. A feature flag that allows a new, more stringent checklist to be tested on a subset of deployments before full rollout is a practical assurance artifact for managing this complexity.

Therefore, advanced implementations can use dynamic, risk-based checklists and should store all audit artifacts in immutable storage to ensure maximum integrity and flexibility.

Related topics

This section focuses on the principles and process of using an automated deployment checkpoint for data workflows. For related topics addressed elsewhere in this book:

- RB-30 rollback capability, see *Chapter 5, Building Scalable Data Pipelines*
- Audit trail and evidence consolidation, see *Chapter 6, Data Governance and Compliance*

Remote and offshore collaboration strategies

Leveraging distributed teams of remote and offshore collaborators is a strategic advantage, offering access to a global talent pool and operational flexibility. This model's success, however, is not automatic. It requires a deliberate framework to overcome the inherent challenges of geographical distance, cultural differences, and asynchronous work. By establishing robust communication protocols, standardized processes, and strategic safeguards, organizations can build highly productive, cohesive, and secure global teams. This approach transforms a potential logistical challenge into a powerful engine for innovation and growth, ensuring that distance does not hinder performance or alignment with business objectives.

Key considerations

Effective management of distributed teams rests on several core principles that must be addressed proactively. The following considerations form the foundation of a successful global collaboration strategy:

- **Communication and trust**: The absence of in-person interaction necessitates explicit communication protocols and intentional trust-building activities. Organizations must define specific channels for different types of communication and establish a regular cadence for meetings to ensure alignment. Teams can accelerate development while clear documentation and regular one-on-one meetings prevent the isolation and misunderstandings that can derail projects. A concrete assurance artifact is a publicly accessible team charter document that outlines communication norms, meeting schedules, and shared expectations.
- **Standardized processes and technology**: Consistency in tools and workflows is critical for maintaining transparency and productivity across different locations. Adopting a unified suite for project management, version control, and knowledge sharing ensures everyone has access to the same information and follows the same procedures. Developers can contribute code efficiently from anywhere and still maintain high-quality standards by using shared CI/CD pipelines. A mandatory pull request template that includes a quality checklist serves as a tangible assurance mechanism.
- **Strategic security and time zone management**: A global footprint introduces complexities related to security and time zone differences that require strategic planning. An asynchronous-first mindset, coupled with designated overlap hours for critical synchronous work, is essential for collaboration without causing burnout. The organization can leverage a global talent pool even as it protects sensitive data by enforcing strict access controls and data protection policies. A time zone-aware on-call rotation schedule provides assurance of continuous operational support.

Therefore, a successful distributed team strategy requires intentional efforts to build trust through clear communication, enforce consistency with standardized tools, and manage global risks with thoughtful security and time zone policies.

Process or steps

Implementing a successful distributed collaboration model follows a structured sequence:

1. **Establish the foundation**: The primary challenge of geographic separation is the potential for ambiguity and team fragmentation. To counter this, organizations must first define clear communication protocols, including designated channels for synchronous and asynchronous interactions, and implement a regular cadence of meetings. Simultaneously, fostering trust through transparency, regular one-on-one check-ins, and virtual team-building activities creates a culture of psychological safety, resulting in a cohesive and aligned team.

2. **Standardize tools and workflows**: Inconsistent processes lead to inefficiency and a lack of visibility, which are magnified in a distributed setting. The next step is to deploy a standardized toolchain for project management (e.g., Jira, Asana), version control (e.g., GitHub, GitLab), and knowledge management (e.g., Confluence). Integrating these tools into a unified workflow, such as a CI/CD pipeline shown in *Figure 8.1*, automates testing and deployment, providing a transparent and efficient development lifecycle that moves forward without compromising on security.

3. **Implement global safeguards**: Operating across different regions introduces risks related to time zones and security. The final step is to implement strategic safeguards, including an asynchronous-first work policy supplemented by a few core overlap hours for essential meetings. To mitigate security risks, enforce the use of **virtual private networks** (**VPNs**), **multi-factor authentication** (**MFA**), and the principle of least privilege for all systems. This ensures secure, continuous progress and protects both company and customer data, with a feature flag system providing assurance through the ability to instantly roll back changes.

Therefore, the process involves first building a foundation of trust and communication, then standardizing the tools and workflows for efficiency, and finally implementing security and time-zone strategies to enable safe, continuous global operations.

Concrete example

A technology company with engineering teams in San Francisco (USA), Berlin (Germany), and Bangalore (India) provides a practical illustration of these principles. The 9.5-hour time difference between San Francisco and Bangalore makes real-time collaboration challenging. To manage this, the company establishes a two-hour daily overlap window for critical meetings, like daily stand-ups and planning sessions. All other communication is handled asynchronously through dedicated Slack channels and detailed task descriptions in Jira.

To maintain code quality and development velocity, the company uses GitHub with a mandatory pull request and code review process. Every code commit automatically triggers a CI/CD pipeline, which runs a suite of automated tests before deploying changes, a process visually represented in *Figure 8.1*. This allows the team to release features daily and still ensure stability through automated canary deployments. For security, all engineers must connect through a company VPN and use MFA to access internal resources. Access to production systems is strictly controlled and logged, with an audit trail link automatically posted to a security-focused Slack channel upon every production deployment, providing clear assurance.

Therefore, a real-world example shows a global team using defined overlap hours, asynchronous tools like Jira and Slack, and an automated CI/CD pipeline to achieve continuous, secure development across multiple time zones.

Cautions and limits

While a distributed model offers significant benefits, organizations must be aware of potential challenges and limitations. A primary risk is team member burnout resulting from pressure to accommodate inconvenient time zones; this can be mitigated by rotating meeting times and promoting a culture that respects local working hours. Cultural misunderstandings can also arise, leading to friction and miscommunication. Proactive cultural awareness training and clear, jargon-free communication standards are essential to bridge these gaps. Furthermore, security vulnerabilities increase with a distributed workforce, as data is accessed from various networks and locations.

These challenges are manageable with deliberate effort. Organizations can scale their global teams without creating security vulnerabilities by implementing and enforcing strict device management policies. For instance, a mandatory security agent installed on all company-issued laptops can enforce endpoint policies, providing assurance that devices meet security standards before connecting to the network. Addressing these limits head-on is crucial for long-term success and prevents small issues from escalating into major obstacles.

Therefore, key risks like burnout, cultural friction, and security vulnerabilities must be actively managed through fair scheduling, cultural training, and strict endpoint security policies to ensure the distributed model remains sustainable.

Glossary

The following are the key terms that have been discussed in this section:

- **Asynchronous communication**: Communication that does not happen in real-time. Participants can send and receive messages at their convenience (e.g., email, documentation comments).
- **CI/CD**: A set of practices and automated workflows that allow development teams to build, test, and deploy code changes more frequently and reliably.
- **MFA**: A security process that requires users to provide two or more verification factors to gain access to a resource, such as a password and a code from a mobile app.

- **PII**: Any data that could be used to identify a specific individual, such as a name, address, or social security number. Strict policies are needed to protect PII.
- **VPN**: A secure, encrypted connection over a public network. It allows remote users to safely access an organization's internal network and resources.

Expert notes

For mature distributed teams, optimizing for speed and quality involves embedding governance directly into automated workflows. Instead of relying on manual checks, high-performing teams automate their quality and compliance verification within the CI/CD pipeline. This can be structured as an automated checklist that the system evaluates before promoting code to the next environment, as exemplified by the logic in *Table 8.1*.

This approach allows teams to accelerate their release cycles while ensuring every deployment meets stringent compliance and quality standards. A critical component of this model is maintaining a robust, asynchronous record of all significant decisions. By mandating that all major architectural or product decisions are recorded in a structured format, such as a decision log file, teams create an immutable audit trail. This log provides context for new team members and prevents the need to revisit settled debates, ensuring that progress can continue uninterrupted even as team composition changes. This decision log file, stored in each service repository, becomes the definitive assurance artifact for architectural governance.

Therefore, advanced teams enhance distributed collaboration by automating quality checks directly within the CI/CD pipeline and maintaining an immutable, asynchronous decision file for all major changes.

Related topics

This section provides a framework for the operational and technical aspects of managing distributed software development teams. For related topics addressed elsewhere in this book:

- DevOps culture and collaboration, see the *DevOps culture and mindset for data teams* section in this chapter
- Pipeline governance and ownership, see the *CI/CD pipelines and automated testing* section in this chapter
- Cross-cultural negotiation tactics or region-specific business etiquette

Lab: DevOps and collaboration on a tiny CI/CD pipeline

In this lab, you will validate foundational CI/CD governance by checking a small pipeline definition for guards (automated checks) and ownership (designated responsible teams) at every stage. You will run a governance check that verifies each stage has the minimum required infrastructure, then intentionally break coverage to observe how the check detects gaps.

Lab environment: All labs run in GitHub Codespaces with a dev container. This lab demonstrates foundational pipeline governance, ensuring every deployment stage has guards and clear ownership.

This lab produces `labs/ch08/artifacts/result.json` containing governance check status, stage-by-stage guard and owner coverage, and accept/reject decisions based on whether every stage meets the minimum infrastructure requirements.

Note: **The lab uses accept/reject for simplicity. In the book text, we use Go/Retry/Kill to describe the same decision workflow.**

Conclusion

This chapter equipped you with essential skills for building and operating reliable CI/CD pipelines for data engineering workloads. You explored the complete deployment workflow, from establishing foundational infrastructure (guards and ownership at every stage) to implementing sophisticated quality checks (data quality validation, budget compliance, rollback capability, audit trail verification) and making informed deployment decisions (Go/Retry/Kill recorded in immutable audit logs).

Through the hands-on lab, you validated the foundational infrastructure that makes reliable deployments possible: ensuring every pipeline stage has automated checks and designated owner teams. This foundation is essential; without it, no amount of sophisticated quality gates can ensure production stability.

Beyond technical automation, you learned how DevOps culture and collaboration practices enable high-velocity deployments: fostering continuous learning and improvement, enabling effective collaboration across distributed and offshore teams, and building shared ownership of production quality. These cultural practices determine whether your CI/CD automation accelerates delivery or merely automates chaos.

These deployment practices become even more vital as your systems scale and evolve. The next chapter explores how to apply these operational principles within complex cloud environments, examining the nuances of transitioning existing data engineering workflows to the cloud and managing hybrid architectures effectively while maintaining the same standards of reliability and auditability.

With automated deployment quality gates in place, Fusion-Mart's data engineering team can now ship updates rapidly while maintaining governance standards.

CHAPTER 9
Cloud Migration and Coexistence Strategies

Introduction

How do you migrate to a new platform without disrupting ongoing operations? Navigating complex organizational changes, especially large-scale cloud migrations, demands robust DevOps practices combined with rigorous governance standards to ensure success and minimize disruption. This chapter builds directly on the foundations of *Chapter 8, DevOps and CI/CD for Data Engineering*, demonstrating how to apply CI/CD automation and evidence-based validation to the specific challenges of cloud migration and coexistence. Whether migrating core business data, transitioning legacy systems, or establishing hybrid architectures for gradual cutover, success depends on combining automated safety checks with proven deployment patterns like blue-green and canary strategies.

Structure

This chapter covers key strategies for successful cloud migration and coexistence:

- Phased and parallel migration approaches
- Minimizing downtime and ensuring business continuity
- Risk mitigation
- Lab: Cloud migration and coexistence

Objectives

By the end of this chapter, you will be able to design phased migration strategies that minimize risk by breaking large migrations into manageable increments with clear rollback points and implement automated validation checks that verify table presence and data consistency across environments before production cutover. You will apply blue-green and canary deployment patterns to enable zero-downtime migrations with gradual traffic shifts and instant rollback, while you establish hybrid coexistence architectures that support legacy and modern systems operating in parallel during transition periods. Finally, you will make informed Go/Retry/Kill decisions at each migration phase using evidence-based validation without compromising integrity.

Phased and parallel migration approaches

Modernizing a legacy data warehouse by migrating to a cloud lakehouse is a significant strategic initiative that extends beyond a simple technology upgrade. This process requires a deliberate approach to manage inherent risks, control costs, and minimize disruption to business operations. Organizations must choose between two primary strategies: a phased, incremental migration or a parallel-run approach culminating in a single cutover. The selection of the optimal path depends on a careful, evidence-based analysis of the organization's risk tolerance, technical capabilities, and business priorities, ensuring the modernization effort delivers value without compromising stability.

Key considerations

Choosing the right migration strategy requires balancing several critical factors. The following points outline the primary considerations for this decision-making process:

- **Risk mitigation vs. speed to market**: A phased migration inherently reduces risk by breaking the project into smaller, manageable units, allowing teams to learn and adapt even as they make progress. However, this approach extends the overall project timeline and introduces integration risks from the extended coexistence period, maintaining two active systems simultaneously increases complexity and potential points of failure. A parallel migration accelerates the timeline but concentrates risk into a single, high-stakes cutover event, which can be managed with a detailed rollback plan.

- **Cost distribution and financial planning**: The financial models for each strategy differ significantly. A phased approach spreads costs over a longer period, aligning with operational budgets, while a parallel strategy requires a larger upfront capital investment to support duplicate infrastructure. Organizations can proceed with a parallel run without impacting the legacy system's budget by using separate, project-based cost tracking codes. Modern FinOps practices enable real-time cost monitoring and allocation during the migration, ensuring budget adherence across both environments.

- **Operational complexity and data consistency**: Maintaining two systems introduces complexity, regardless of the strategy. A phased migration requires managing data synchronization between the legacy and new systems for an extended duration. A parallel migration demands robust, real-time data replication, but for a shorter period, and still requires automated validation scripts to ensure data integrity before the final cutover.

Therefore, the decision between a phased and parallel migration hinges on a strategic evaluation of the organization's tolerance for risk, its financial model, and its technical capacity to manage data consistency.

Process or steps

Selecting and executing a migration strategy involves three key steps:

1. **Analyze historical performance**: To ground the decision in evidence, the project team must analyze past performance on complex initiatives. By reviewing artifacts like the decision log and performance metrics, combining quantitative measures (change failure rate, deployment frequency, mean time to recovery) with qualitative insights from post-mortem reviews, the team identifies historical process bottlenecks and assesses the organization's capacity for managing large-scale change, which directly informs the risk assessment for each migration strategy.

2. **Define migration strategy and plan**: Based on the analysis, the team selects either a phased or parallel approach and documents it in the migration plan document. This plan details the scope, sequence of events, resource allocation, validation criteria, and a comprehensive rollback procedure, ensuring all stakeholders have a clear and actionable roadmap, while a feature flag system allows for immediate reversion if needed. The plan should be maintained in version control with pull-request-based governance, ensuring all changes receive peer review and approval before implementation.

3. **Implement and validate with a controlled cutover**: The team executes the migration according to the plan, using a controlled cutover mechanism to minimize business disruption. For a parallel migration, this often involves a blue-green deployment, as shown in *Figure 9.1*, where traffic is shifted to the new system only after it passes all validation checks, ensuring business continuity is maintained throughout the transition.

Therefore, an effective migration follows a process of analyzing past performance to inform the strategy, documenting the plan in detail, and executing a controlled cutover using proven patterns to ensure a successful outcome.

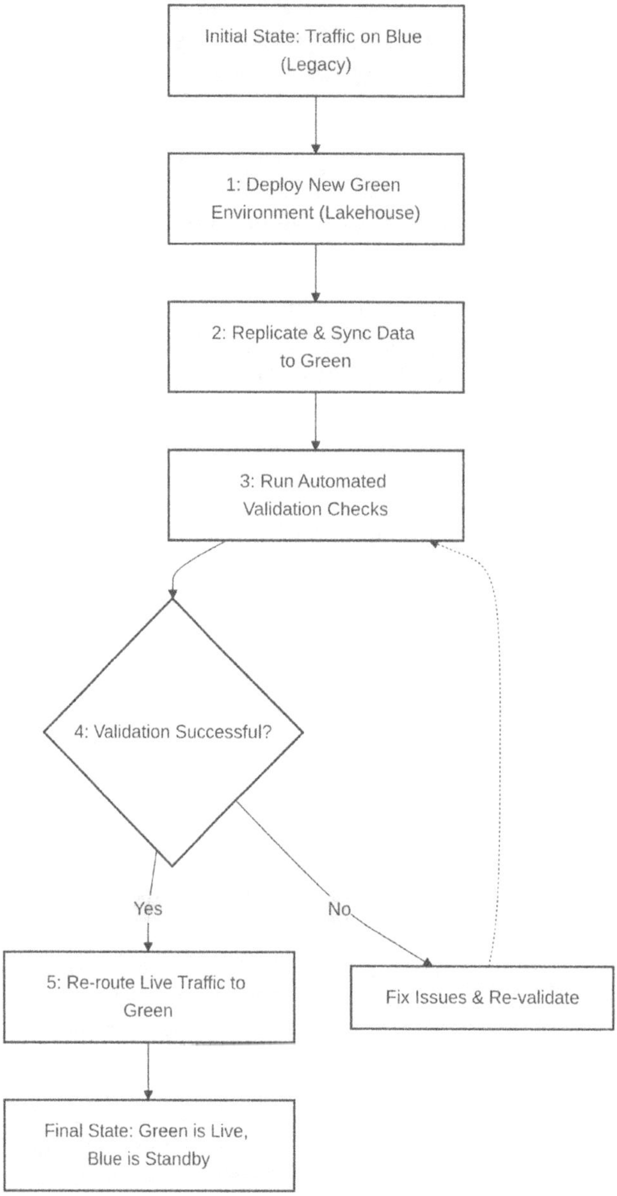

***Figure 9.1**: Steps to minimize downtime and risk using a blue-green deployment strategy*

Concrete example

Consider Fusion-Mart, a retail company migrating its on-premise data warehouse to a cloud lakehouse. After analyzing its historical performance data, the leadership team discovers that past large-scale, single-weekend deployments have consistently resulted in significant operational disruptions. This evidence points toward a lower risk appetite, making a phased migration the more prudent choice. They decide to start with a single data domain: Sales. The team migrates the Sales data and associated analytical workloads to the new lakehouse first. For a period, they run reports from both the old and new systems, using an automated dashboard, to compare results and validate data integrity. Once confident, they switch only the sales analysts to the new platform. This incremental success builds confidence and provides valuable lessons for migrating the next domain, Inventory, while the rest of the business continues to operate without interruption, all tracked via a central change management log.

Therefore, Fusion-Mart used historical performance data to choose a phased migration, starting with the Sales domain to validate the process and minimize business risk before proceeding with other domains.

Cautions and limits

While structured migration strategies provide a clear path, organizations must be aware of potential challenges and prerequisites. A phased migration can suffer from scope creep or migration fatigue if the phases are not clearly defined and time-boxed, potentially extending the timeline indefinitely. Conversely, a parallel migration's primary risk is the complexity of the final cutover; a failure at this stage can cause significant, widespread business disruption. A critical prerequisite for either approach is having personnel skilled in both legacy system maintenance and modern cloud data platforms. Organizations can address this skills gap through targeted training programs and still meet project deadlines. To mitigate cutover risk, teams must implement a thoroughly tested rollback mechanism, such as the blue-green strategy detailed in *Figure 9.1*, which is confirmed operational via a mandatory pre-flight check before the event. Phased migrations also risk creating zombie architectures, legacy systems that remain partially active due to unresolved dependencies, accumulated technical debt, or stakeholder reluctance, making full decommissioning difficult despite ongoing maintenance costs.

Therefore, both phased and parallel migrations have distinct risks, prolonged complexity versus cutover failure, and both require careful planning and skilled personnel to succeed.

Glossary

This section defines key terms used throughout the discussion:

- **Cloud lakehouse**: A modern data architecture that combines the low-cost, flexible storage of a data lake with the data management and query features of a data warehouse.

- **Blue-green deployment**: A continuous deployment strategy that reduces downtime and risk by running two identical production environments, referred to as blue and green. Only one environment is live at any given time, serving all production traffic.

- **Cutover**: The decisive event in a migration process where the new system is made live, and the legacy system is decommissioned or relegated to a read-only state.
- **Data domain**: A logical grouping of data that pertains to a specific business area or function, such as Customer, Sales, or Inventory.
- **Stateful application**: An application that stores data about a client session or transaction history. The application's response to a request depends on this stored state, making rollbacks more complex than for stateless applications.

Expert notes

Experienced practitioners often employ hybrid strategies to optimize for specific circumstances. For instance, a team might use a phased approach to migrate data domains incrementally into the cloud lakehouse. Once all data is migrated and validated, they can then use a parallel-run strategy for a short period to switch over to the primary analytical applications. This combines the risk mitigation of the phased approach for the data with the speed of the parallel approach for the applications. Furthermore, advanced teams heavily automate data validation and reconciliation between the two systems, using checksums and record counts to ensure consistency without manual intervention. This allows them to maintain data integrity even as they accelerate the migration timeline, with all validation results automatically logged to an immutable audit trail for compliance.

Therefore, experts often create hybrid migration models and use extensive automation for data validation to balance risk, speed, and data integrity effectively.

Related topics

This section focuses specifically on the strategic choice between phased and parallel data warehouse migration strategies. For related topics addressed elsewhere in this book:

- CI/CD automation and deployment patterns, see *Chapter 8, DevOps and CI/CD for Data Engineering*
- Data pipeline construction, see *Chapter 5, Building Scalable Data Pipelines*

Minimizing downtime and ensuring business continuity

Maintaining uninterrupted business operations during a data platform migration is a critical imperative. Any significant downtime can cause substantial financial loss, damage customer trust, and disrupt core functions. To manage this risk, organizations can adopt advanced strategies that allow for complex system changes, such as a cloud migration, with minimal impact on users. By implementing parallel environments and controlled transitions, teams can thoroughly validate a new platform before it handles production traffic. These strategies enable rapid innovation while ensuring the stability of core business functions, with readiness confirmed by a validation report. This approach transforms a high-risk event into a managed, evidence-based process that ensures business continuity.

Key considerations

Here are the most important points to consider when planning a data platform migration:

- The primary objective is to protect business operations from disruption. A migration is not merely a technical task; it is a business-critical procedure where any service interruption can lead to immediate revenue loss and erode customer confidence. Therefore, the migration plan must prioritize continuity, ensuring that progress on the technical front proceeds without compromising the stability of the live production environment, which can be tracked via shared monitoring dashboards.

- Blue-green deployments provide a powerful framework for minimizing risk. This strategy involves maintaining two identical production environments, allowing the new platform (green) to be fully tested and validated with real data before any user traffic is directed to it. This allows teams to validate the new system under production load without impacting live users, using a feature flag to control traffic routing for a safe cutover.

- Data consistency between the old and new systems is non-negotiable. Without reliable data synchronization, a cutover could result in data loss or corruption, undermining the entire migration effort. Employing techniques like **change data capture** (**CDC**) ensures that the new platform is always up to date, enabling a seamless transition and still maintaining data integrity throughout the process.

Therefore, a successful migration hinges on using parallel environments like blue-green deployments and robust data synchronization to minimize downtime and provide rapid rollback capabilities.

Process or steps

Executing a near-zero downtime migration using a blue-green strategy involves four key steps, as illustrated in the deployment flowchart in *Figure 9.1*:

1. **Provision and configure the green environment**: The initial cause is the need for a new data platform. The action is to build out the new environment (green) as an identical, parallel system to the existing production environment (blue). This includes provisioning infrastructure, installing the new data platform software, and configuring all necessary networking and security controls, with the final configuration stored in a version-controlled `terraform_state_file`.

2. **Synchronize data continuously**: To ensure data consistency, the next action is to establish a continuous data synchronization pipeline from the blue environment to the green one. This is often achieved using CDC, which identifies and replicates all data modifications (inserts, updates, deletes) in near real-time. The result is a fully hydrated green environment that mirrors the production data of the blue environment, allowing the migration to proceed methodically even as the legacy system continues to serve live traffic. Note that CDC has limitations: it struggles with legacy flat-file systems lacking change tracking, batch-oriented sources, and databases without transaction logs. Alternatives include timestamp-based incremental loads or full-table comparisons with checksums.

3. **Validate and switch traffic**: Once the green environment is stable and data is synchronized, the cause for action is the readiness for cutover. The action is to redirect a portion or all of the user traffic from the blue environment to the green one, typically by updating a **Domain Name System** (**DNS**) record or a load balancer configuration. The result is that users begin interacting with the new platform with no perceptible service interruption.

4. **Monitor and decommission**: After the traffic switch, the final action is to closely monitor the green environment's performance and stability. If any critical issues arise, traffic can be immediately reverted to the blue environment. Once the new system is confirmed to be operating correctly over a predefined period, the blue environment can be decommissioned to save costs, completing the migration.

Therefore, the process involves preparing a parallel environment, synchronizing data using techniques like CDC, redirecting traffic in a controlled manner, and decommissioning the old system only after validating the new one.

Concrete example

An e-commerce company needs to migrate its orders database from a self-hosted, legacy SQL server to a managed cloud database service to improve scalability and reduce operational overhead. To avoid disrupting sales, the team adopts a blue-green deployment strategy. First, they provision the new cloud database (the green environment) and configure it to match the production schema. They then use a CDC tool to perform an initial bulk load of all existing order data from the old server (blue) to the new one. After the initial load, the CDC tool continuously streams all new transactions, such as new orders, status updates, and cancellations, to the green database in real-time. The team could test the new database's performance with production data and still maintain operational stability on the old system, using an API gateway configuration to manage the traffic switch. Once testing confirms the green database is stable and performant, they update their application's connection string to point to the new cloud database during a low-traffic maintenance window. The switch is instantaneous for users. The old database server remains online as a hot standby for 72 hours before being decommissioned.

Therefore, an e-commerce company uses CDC to synchronize its **`orders`** database to a new cloud platform, allowing it to switch over application traffic with zero-downtime and an immediate rollback option.

Cautions and limits

While blue-green deployments are highly effective, they come with challenges that require careful management. The most significant limitation is the cost and complexity of maintaining two full production environments, which can double infrastructure expenses and increase operational overhead during the migration period. Furthermore, ensuring perfect data synchronization can be difficult, especially with complex database schemas or high transaction volumes; subtle bugs in the synchronization logic could lead to data inconsistencies. Finally, stateful applications that write data to both environments simultaneously (a split-brain scenario) must be avoided, as this can lead to data corruption that is difficult to reconcile. Organizations can mitigate this by implementing automated data reconciliation scripts while running both systems, ensuring data integrity is confirmed by an audit trail before decommissioning the old environment.

Therefore, key limitations include the high cost of duplicate infrastructure and the technical complexity of ensuring flawless, real-time data synchronization without introducing inconsistencies.

Glossary

This section defines key terms used throughout the discussion:

- **CDC**: A set of software design patterns used to determine and track the data that has changed so that action can be taken using the changed data. It is a common method for ensuring data remains synchronized between two systems in near real-time.
- **Data reconciliation**: The process of comparing data between two sources (e.g., the blue and green environments) to identify and correct any discrepancies. This is a critical step to ensure data integrity before completing a migration.

Expert notes

For highly critical systems, the traffic switch can be executed more granularly than an all-or-nothing cutover. Using a technique called canary releasing, a small percentage of user traffic (e.g., 1%) is initially routed to the green environment. This allows teams to monitor the new platform's performance with a limited blast radius. If monitoring dashboards show no errors, traffic can be incrementally increased until 100% is on the new system. This allows for incremental validation without committing to a full cutover, where the rollback mechanism is a simple configuration change in the service mesh. Additionally, handling database schema changes during a migration requires careful planning. Schema updates must be backward-compatible to ensure the application can work with either the blue or green database during the transition period. This dual-readiness is often managed through a feature flag configuration that controls application behavior based on which database is active.

Therefore, advanced techniques like canary releasing allow for a gradual traffic shift to de-risk the cutover, while backward-compatible schema changes are essential for maintaining application functionality during the transition.

Related topics

This section focuses specifically on strategies for minimizing downtime during a data platform migration. For related topics addressed elsewhere in this book:

- Blue-green and canary deployment strategies, see *Chapter 8, DevOps and CI/CD for Data Engineering*
- Rollback capability (RB-30), see *Chapter 5, Building Scalable Data Pipelines*

Risk mitigation

Robust fallback and comprehensive testing protocols are essential governance controls that support evidence-based decision-making at critical migration gates. In high-stakes cloud migrations, these practices provide the evidence needed to make informed Go/Retry/Kill decisions while preparing the safety net required for safe rollback when issues are detected. Testing protocols generate the validation evidence, while fallback procedures enable confident kill decisions when necessary. Establishing clear, well-rehearsed rollback strategies provides a vital safety net, empowering teams to proceed with confidence. These decisions map directly to automated pipeline gates: Go allows the migration phase to proceed, Retry triggers remediation and re-validation, and Kill halts the migration and activates rollback.

Key considerations

When designing a strategy for complex change, several core principles must be considered to balance progress with stability:

- **Risk management alignment**: Robust fallback and testing are not merely technical tasks; they are fundamental risk management instruments. They provide a safety net that enables confident execution, allowing teams to pursue significant system improvements without exposing the business to unacceptable failure modes. The existence of a version-controlled disaster recovery plan provides documented assurance that recovery procedures are defined and understood.
- **Proactive strategy design**: Effective recovery paths must be designed into the migration strategy from the outset, not added as an afterthought. A plan that includes a clear,

tested rollback procedure is inherently more resilient and trustworthy. This approach allows teams to innovate on deployment strategies and still maintain operational stability, using a versioned infrastructure as code commit as a tangible recovery point.

- **Business continuity focus**: The ultimate goal of these controls is to ensure uninterrupted business operations during major transitions. Implementing patterns like blue-green deployments exemplifies this principle, ensuring services remain available even as underlying components are upgraded. This allows for progress on feature delivery without compromising the user experience, where a load balancer configuration that can instantly switch traffic serves as a key assurance artifact.

Therefore, fallback and testing should be treated as proactive design principles for risk management and business continuity, not as reactive measures for disaster recovery.

Process or steps

Implementing a reliable fallback and testing protocol for a complex migration involves three critical stages:

- **Establish baselines and success criteria**: Before initiating any change, capture the performance metrics and functional correctness of the current system. Define what a successful migration looks like in measurable terms, such as specific latency targets, error rate thresholds, and **central processing unit** (**CPU**) utilization. This creates an objective standard for validating the new system and confirming the health of the old one if a rollback is necessary.

- **Execute pre- and post-migration testing**: Conduct a comprehensive suite of tests, both synthetic tests (scripted scenarios covering known use cases) and replay-based tests (captured production traffic replayed against the new environment), including functional, performance, and integration tests in a staging environment that closely mirrors production. After the migration, run the same test suite against the new production environment to verify that the predefined success criteria are met. This process validates the new system's integrity and catches issues before they impact users, even as the team prepares to proceed with the production deployment, retaining the ability to revert via a rollback script.

- **Activate the rollback plan**: If post-migration testing reveals a critical failure, execute the pre-defined and rehearsed procedure to revert to the last known stable state. This action, which could involve redirecting network traffic, restoring a database snapshot, or redeploying a previous application version, minimizes the **mean time to recovery** (**MTTR**) and contains the business impact of the failure.

Therefore, a successful process hinges on defining success upfront with clear metrics, testing rigorously against that definition, and having a practiced, automated plan to revert if validation fails.

Concrete example

A team migrating a monolithic e-commerce application's payment service to a new cloud-native microservice provides a practical illustration. They adopt a blue-green deployment strategy, as shown in *Figure 9.1*, to minimize risk. The new microservice (the green environment) is deployed alongside the old monolith (the blue environment). Initially, a load balancer routes a small percentage of internal traffic to the green environment for final validation. After observing stability, the team shifts 10% of live customer traffic. Almost immediately, monitoring dashboards show a spike in payment failures. Because they planned for this contingency, the team was able to test this major architectural change in production without causing a prolonged outage. The on-call engineer executes a single, pre-approved command to update the load balancer via its version-controlled load balancer configuration file, redirecting 100% of traffic back to the stable blue environment. The entire rollback is completed in under one minute, preventing significant revenue loss and customer impact.

Therefore, a blue-green deployment for a critical payment service allowed a team to use live traffic for validation and rapidly roll back upon detecting an issue, preventing a widespread outage.

Cautions and limits

While essential, fallback and testing strategies have inherent limitations that require careful management. Teams can manage stateful migrations and still provide a recovery path, but only by implementing data synchronization strategies and acknowledging system complexity. For instance, rollbacks are significantly more complex for stateful applications where database schemas have changed; a simple code revert does not address data transformations. This risk can be mitigated with patterns like parallel data writes during the transition and a documented procedure for restoring a database from a recent, validated **`database_snapshot`**. Furthermore, no test suite is perfect, and staging environments cannot fully replicate the unpredictability of production traffic. This gap can be addressed with canary releases and robust monitoring to detect unforeseen issues with a limited blast radius. Finally, a rollback plan that is not regularly rehearsed can fail when needed most, so it must be treated as production code and continuously validated.

Therefore, effective fallback strategies require careful planning for data consistency, acknowledging the limits of pre-production testing, and continuously rehearsing rollback procedures to ensure they remain viable.

Glossary

This section defines key terms used throughout the discussion:

- **Canary release**: A technique to reduce the risk of introducing a new software version in production by slowly rolling out the change to a small subset of users before making it available to everybody.

- **MTTR**: The average time it takes to recover from a product or system failure, from the time of failure detection to the moment the system is fully operational again.

Expert notes

Experienced teams elevate fallback strategies beyond simple, binary rollbacks. By using feature flags, they gain granular control to disable specific problematic features within a new release without reverting the entire deployment. This allows the system to move forward even as isolated issues are addressed via a feature flag configuration file. The most reliable rollback plans are also fully automated; by automating rollbacks within the CI/CD pipeline, teams can accelerate their deployment frequency while improving their recovery time objectives. This automation removes the risk of manual error under pressure. Finally, experts recognize that maintaining parallel environments for strategies like blue-green deployment incurs infrastructure costs. This expense is a calculated business decision, weighing the cost of redundancy against the potential cost of downtime for a critical service.

Therefore, experts elevate fallback strategies by using feature flags for granular control, fully automating recovery procedures in CI/CD, and strategically balancing the cost of redundancy against the business impact of downtime.

Related topics

This section focuses on the strategy and process of implementing fallback and testing controls. For related topics addressed elsewhere in this book:

- Evidence-based governance gates, see *Chapter 6, Data Governance and Compliance*
- CI/CD pipeline automation, see *Chapter 8, DevOps and CI/CD for Data Engineering*
- Code-level implementations for data consistency or backward compatibility

Lab: Cloud migration and coexistence

In this lab, you will validate a cloud migration plan by checking table presence and row counts across on-premises and cloud environments, running a migration gate that verifies all planned tables exist and that the migrated data matches.

Lab environment: All labs run in GitHub Codespaces with a dev container. This lab demonstrates foundational migration validation, ensuring table presence and data consistency across on-prem and cloud environments.

This lab produces `labs/ch09/artifacts/result.json` containing migration validation status, table-by-table presence checks, row count comparisons, and accept/reject decisions based on whether the migration plan meets all safety requirements.

Note: **The lab uses accept/reject for simplicity. In the book text, we use Go/Retry/Kill to describe the same decision workflow.**

Conclusion

This chapter equipped you with essential skills for planning and executing safe cloud migrations. You explored the complete migration workflow, from designing phased strategies that enable gradual cutover to implementing automated validation checks that verify data consistency across on-premises and cloud environments, applying proven deployment patterns like blue-green and canary strategies for zero-downtime migration, and using evidence-based governance gates to make informed Go / Retry / Kill decisions at each migration phase.

Through the hands-on lab, you implemented the foundational validation gate that makes safe migrations possible: automated checks for table presence and data integrity across environments. This validation layer is essential; without it, no sophisticated migration strategy can prevent production failures caused by missing data or environment mismatches. The lab demonstrated how automated validation transforms manual verification processes into reliable, repeatable safety checks.

Beyond technical patterns, you learned how DevOps practices enable successful migrations: fostering collaboration between infrastructure and application teams, establishing shared responsibility for migration safety, and building automated feedback loops that detect issues early. These practices determine whether your migration accelerates business transformation or becomes a source of prolonged disruption.

These migration practices and governance standards become even more critical as your data infrastructure grows in complexity. The next chapter explores how to apply operational excellence principles to data warehouse sizing and concurrency planning, examining how to optimize resource allocation and manage workload performance while maintaining the same standards of automation and governance established in this chapter.

Fusion-Mart's phased migration, starting with Sales, then Inventory, validated the process safely while the rest of the business continued without interruption.

CHAPTER 10
Scaling Data Platforms with Optimization

Introduction

How do you scale a data platform without sacrificing performance or breaking the budget? Scaling data platforms effectively requires combining technical optimization with robust DevOps practices and governance standards, without governance, uncontrolled optimizations can introduce instability or hidden cost regressions. Building on the CI/CD automation and cloud migration strategies from previous chapters, this chapter focuses on performance tuning, cost optimization, and observability practices that enable sustainable growth. You will learn to identify and resolve performance bottlenecks, optimize cloud resource costs, implement proactive monitoring with **service level indicators** (**SLIs**), metrics like data freshness for data systems and latency for applications, and apply these techniques through hands-on validation of a promotion gate that ensures data products meet quality standards before production deployment.

SLIs work in conjunction with SLOs to provide a complete observability framework. While SLIs are the quantifiable metrics that measure specific aspects of system performance, such as query latency, data freshness, or pipeline success rates, SLOs define the acceptable thresholds for these indicators. For example, an SLI might measure query execution time, while the corresponding SLO sets the target: *95% of queries must complete within 2 seconds*. Together, SLIs and SLOs enable data teams to make objective, evidence-based decisions about system health and deployment readiness.

Structure

This chapter covers the following topics:

- Data platform performance tuning
- Cloud cost optimization strategies
- Observability practices for proactive monitoring
- Delta Lake performance and compaction
- Emerging data trends
- Lab: Scaling and optimizing tiny warehouses

Objectives

By the end of this chapter, you will be able to identify and resolve performance bottlenecks through systematic analysis using query profiling and performance metrics collection, and optimize cloud costs by implementing right-sizing strategies, reserved capacity, and storage lifecycle policies that balance performance and budget. You will implement proactive monitoring with SLIs that detect degradation before users are impacted, enabling daytime issue resolution, while you apply Delta Lake optimization techniques, including Z-ordering for clustering and automatic compaction for improved query performance. Finally, you will validate data product quality through automated promotion gates that ensure only tested, compliant changes reach production.

Data platform performance tuning

As data volume and velocity grow, sometimes rapidly, sometimes gradually, a data platform's ability to scale efficiently and deliver timely insights becomes paramount for business success. Proactive performance tuning is essential for managing this growth. This involves a holistic strategy that combines three core techniques: optimizing data retrieval with efficient queries, structuring data for scale with partitioning, and reducing latency with intelligent caching. By mastering these interconnected disciplines, organizations can ensure their data platform remains a powerful and responsive engine for analytics and decision-making, transforming massive datasets from a performance liability into a strategic asset.

Key considerations

Here are the most important points to consider when tuning a data platform:

- **Adopt a holistic strategy:** Query optimization, data partitioning, and caching are not isolated solutions but interconnected components of a single performance strategy. For instance, an optimized query will yield limited benefits if it must scan a massive, unpartitioned table. True performance gains come from applying these techniques in concert, allowing teams to improve data access while maintaining a stable and manageable architecture. A version-controlled data contract ensures that schema changes support these tuning efforts.

- **Prioritize based on business impact:** Not all slow queries are created equal; therefore, tuning efforts should be focused where they deliver the most value. Prioritize optimizing systems that directly impact critical business operations, such as customer-facing applications or executive sales dashboards. This approach ensures that technical resources are invested to improve key workflows and decisions, even as the overall system continues to evolve. SLA document helps formalize these performance expectations.

- **Embrace continuous improvement:** Performance tuning is an ongoing process, not a one-time project, as illustrated in the continuous cycle shown in *Figure 10.1*. As data patterns and business requirements change, new bottlenecks will emerge. Establishing a cycle of monitoring, analysis, and iterative refinement allows teams to proactively manage performance without disrupting users. Automated alerting from a unified observability dashboard provides the foundation for this proactive stance.

Therefore, effective data platform tuning requires a holistic, business-aligned, and continuous approach that treats performance as an ongoing process, not a singular fix.

Process or steps

Improving data platform performance involves a systematic, four-step process:

1. **Identify performance bottlenecks:** When a process, like a report or dashboard, runs slowly, the first step is to diagnose the root cause. This involves using monitoring tools and analyzing the query execution plan (a database's roadmap for retrieving data) to pinpoint inefficiencies. Common issues include full table scans on large tables, suboptimal join operations, or resource contention.
2. **Implement a targeted optimization:** With the bottleneck identified, select and apply the most appropriate tuning technique. This could involve rewriting a query to filter data earlier, adding a database index to a frequently queried column, or partitioning a large table by a key like date or region.
3. **Validate and deploy the change:** Before releasing a change to production, validate its effectiveness in a staging environment. This step confirms the performance improvement and ensures the change does not introduce unintended side effects or regressions. Deploying the change with a feature flag provides a crucial safety net, allowing for an instant rollback if necessary.
4. **Monitor and iterate:** After deployment, continuously monitor key performance metrics to confirm the fix is working as expected and has not created new problems elsewhere. This feedback loop is critical, as the results inform the next cycle of performance tuning. A documented rollback procedure ensures that any negative impacts can be swiftly reversed.

Therefore, a structured process of identifying bottlenecks, implementing targeted fixes, validating changes, and monitoring results creates a reliable cycle for performance improvement.

Concrete example

Fusion-Mart's quarterly sales report, which queries a multi-billion row **`sales_transactions`** table, began taking several hours to complete, threatening to delay critical business planning. Following the performance tuning process, the data team analyzed the query's execution plan and discovered it was performing a full table scan, reading every record just to filter for transactions within the last quarter. The cause was clear: the query was inefficiently structured for such a large dataset. To resolve this, the team implemented a data partitioning strategy on the **`sales_transactions`** table, dividing it into monthly segments based on the **`transaction_date`** column. This change allows the query to scan only the three relevant monthly partitions instead of the entire table. The change was validated in a staging environment, where the report's runtime dropped to under five minutes. After a successful deployment managed via a configuration flag, the team's monitoring dashboards confirmed the sustained performance gain, enabling faster, more agile business decisions while the underlying data continued to grow.

Therefore, by partitioning its massive sales table by date, Fusion-Mart reduced a multi-hour report query to minutes, directly illustrating how structural changes solve performance bottlenecks.

Cautions and limits

While performance tuning techniques are powerful, they come with potential risks that must be managed to prevent unintended consequences. For example, over-indexing a table can accelerate read queries but significantly slow down write operations like **`INSERT`** and **`UPDATE`**, as every index must be updated with each data modification. Similarly, choosing an incorrect partitioning key can lead to data skew (unevenly sized partitions), which negates the performance benefits. Caching strategies introduce their own challenge: cache invalidation. If not managed carefully, applications can serve stale data from the cache, leading to incorrect reports and poor user experiences. Teams can mitigate these risks by implementing changes incrementally and still achieve their performance goals. A mandatory, automated test suite that runs before deployment serves as a critical check against regressions. Additionally, ongoing maintenance overhead, such as index rebuilds and partition management, adds operational cost that must be factored into the optimization strategy.

Therefore, performance tuning requires careful management of risks like over-indexing, poor partitioning choices, and stale cache data to avoid creating new problems.

Glossary

Here are definitions for key terms used in this section:

- **Query execution plan:** A sequence of steps used by a database to access data for a SQL query. Analyzing this plan is crucial for identifying performance bottlenecks.
- **Indexing:** A database feature that creates a special lookup table to speed up data retrieval. It is analogous to the index in the back of a book.
- **Data partitioning:** The process of dividing a large logical table into smaller, more manageable physical segments based on a specific key, such as date or region.

- **Caching:** The practice of storing copies of frequently accessed data in a high-speed, temporary storage layer (like memory) to reduce latency and database load.
- **Cache invalidation:** The process of marking or removing data in a cache when the original data source has changed, ensuring data freshness.
- **Data skew:** A condition in distributed systems where data is unevenly distributed across different partitions or nodes, often leading to performance bottlenecks.

Understanding these fundamental concepts enables engineers to make informed decisions and enhance system performance without introducing unnecessary architectural complexity. A centrally managed data dictionary provides a single source of truth for all technical terms.

Expert notes

For experienced practitioners seeking to further optimize data platforms, several advanced techniques offer significant performance advantages. Materialized views, which pre-compute and store the results of complex and frequently executed aggregation queries, can provide sub-second response times for dashboards that would otherwise take minutes to load. Modern query engines also feature **adaptive query execution** (**AQE**), a capability that allows the engine to adjust query plans on the fly based on real-time data statistics, automatically mitigating issues like data skew in joins. Implementing a multi-tiered caching architecture, combining a local in-process cache for instant access with a larger, shared distributed cache, can deliver optimal performance even as application traffic scales. Mastering these advanced methods is a key differentiator for senior data professionals. Access to these advanced features is governed by a change advisory board to ensure they are implemented responsibly.

Therefore, advanced techniques like materialized views, adaptive query execution, and multi-tiered caching offer expert-level control over platform performance and scalability.

Related topics

This section focuses on logical and structural performance tuning techniques. For related topics addressed elsewhere in this book:

- Data pipeline optimization, see *Chapter 5, Building Scalable Data Pipelines*
- Query patterns and data contracts, see *Chapter 3, Data Integration Strategy, Business Impact, and ROI*

Cloud cost optimization strategies

Effective cloud cost optimization is a critical discipline for managing data products and aligning technology investments with business objectives. For an organization like Fusion-Mart, it involves implementing concrete, reversible improvements that reduce expenditure while preserving the performance and reliability of their data platforms. This requires a systematic approach that moves beyond simple cost-cutting to intelligent resource management. By adopting strategies such as right-sizing and leveraging flexible compute options, Fusion-Mart can achieve significant savings and still deliver high-quality data services. The entire process is governed by a commitment to controlled implementation, with every change validated through a version-controlled deployment pipeline.

Key considerations

Before implementing specific cost-saving tactics, it is essential to establish a strategic framework. The following points are central to a successful cloud cost optimization initiative:

- **Align optimization with business goals:** Cloud cost optimization should be treated as a core component of data product design, not an afterthought. Each proposed change must be a reversible improvement that supports Fusion-Mart's budget targets without negatively impacting service levels. This alignment ensures that technical teams can pursue efficiency gains and still meet their performance commitments, with changes tracked via a mandatory change request ticket in a ticketing system like Jira, ServiceNow, Azure DevOps, etc.
- **Embrace proactive resource management:** The foundation of cost efficiency is preventing waste before it occurs. Strategies like resource right-sizing, which match infrastructure to actual workload demand, are fundamental. This proactive stance avoids the common pitfall of over-provisioning and ensures that data products are built for efficiency from the design phase, even as new features are added. A **configuration management database** (**CMDB**) provides a clear record of all resource adjustments.
- **Leverage cloud-native pricing models:** Cloud providers offer flexible pricing models that can yield substantial savings for the right workloads. Leveraging spot instances for fault-tolerant, non-critical processes is a powerful, reversible optimization. This allows Fusion-Mart to reduce compute costs for tasks like batch data processing, knowing that the change can be easily rolled back using a feature flag if performance is affected.

Therefore, a successful cost optimization strategy aligns technical changes with financial goals, focuses on proactive resource management, and leverages flexible pricing models as reversible improvements.

Process or steps

Implementing a sustainable cost optimization practice involves a continuous cycle of analysis, action, and validation. This iterative process, as shown in *Figure 10.1*, ensures that improvements are data-driven and their impacts are carefully measured:

1. **Monitor and analyze usage:** The first step is to gain deep visibility into current cloud consumption. By using native cloud monitoring tools and a well-defined tagging

strategy, teams collect detailed usage data on resources like virtual machines and databases. This data reveals underutilized assets, identifies peak usage patterns, and uncovers anomalies, providing a clear basis for action while an audit trail logs all access to the monitoring dashboards.

2. **Implement reversible changes:** With clear opportunities identified, the next step is to implement concrete, reversible changes. This could involve downsizing an over-provisioned database, re-architecting a batch process to run on spot instances, or scheduling development environments to shut down outside of business hours. Each change is deployed through a **continuous integration/continuous deployment (CI/CD)** pipeline, allowing for automated, low-risk implementation and simple rollback if needed.
3. **Validate and iterate:** After a change is deployed, its impact must be validated. Teams must monitor both cost dashboards and application performance metrics to confirm that savings were achieved without introducing new problems. This feedback loop provides the crucial data needed to justify the change, inform the next cycle of optimizations, and continuously refine the financial efficiency of the data product. Define specific rollback criteria upfront, such as a 10% increase in error rates or response times exceeding SLO thresholds, to enable rapid, objective revert decisions.

Therefore, a systematic cost optimization process relies on a continuous loop of monitoring usage, implementing reversible changes through automated pipelines, and validating the financial and performance outcomes.

The following figure shows the continuous improvement cycle for a data platform:

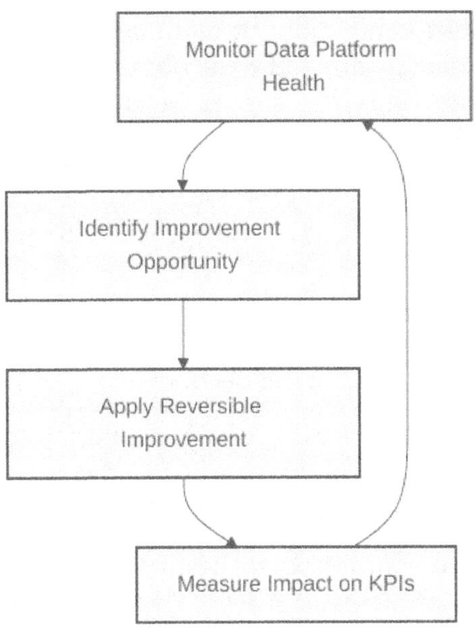

***Figure 10.1**: The continuous improvement cycle for a data platform*

Concrete example

Consider Fusion-Mart's nightly **extract, transform, load** (**ETL**) pipeline, which processes sales data. Initially, the pipeline ran on a large, expensive `m5.4xlarge` on-demand instance that was provisioned to handle the highest potential load, but monitoring data revealed it was only busy for four hours each night. The team identified this as a prime candidate for optimization. They implemented a reversible change by containerizing the ETL application and deploying it on a Kubernetes cluster configured to use AWS Spot Instances. This allowed them to achieve significant cost reductions and still complete the processing within the required window. The entire change was controlled by a feature flag, allowing them to instantly revert to on-demand instances if the spot instance interruption rate became too high.

Therefore, by switching a nightly ETL job from an oversized on-demand instance to a containerized workload on spot instances, Fusion-Mart drastically cut costs without impacting data delivery.

Cautions and limits

While cloud cost optimization offers significant benefits, it also presents challenges that require careful management. A primary risk is that aggressive right-sizing can lead to under-provisioning, causing performance degradation that impacts end-users. To mitigate this, changes should be implemented incrementally, allowing teams to validate performance at each stage without disrupting service. Another limitation involves spot instances; their interruptible nature makes them unsuitable for critical, stateful applications that cannot tolerate downtime. Therefore, teams must carefully architect workloads for fault tolerance before deploying them on spot capacity. Finally, a robust tagging strategy is a prerequisite for accurate cost allocation, but maintaining it requires disciplined governance. Without it, cost visibility diminishes, making it difficult to attribute spending and identify new savings opportunities. All optimization activities should be documented in a runbook to ensure procedures are followed consistently.

Therefore, teams must balance cost savings with performance, use spot instances only for appropriate workloads, and maintain a disciplined tagging strategy to avoid unintended consequences.

Glossary

Here are definitions for key terms used in this section:

- **Resource right-sizing:** The process of analyzing performance and capacity metrics of a cloud resource and adjusting its size (e.g., CPU, memory, storage) to match the actual workload demand, eliminating waste from over-provisioning.
- **Spot instances:** Unused compute capacity that cloud providers offer at a significant discount compared to on-demand prices. The provider can reclaim these instances with short notice, making them ideal for fault-tolerant and flexible workloads.

- **Observability loop:** A continuous cycle where a system's outputs are monitored and analyzed to generate insights that inform future improvements to the system's performance, reliability, or efficiency.
- **Tagging strategy:** The practice of applying metadata labels (key-value pairs called tags) to cloud resources. This enables organizations to categorize resources by project, owner, or environment for cost tracking, automation, and access control.

Expert notes

For advanced practitioners, cost optimization can be integrated more deeply into engineering workflows. By defining infrastructure configurations in code using tools like Terraform, teams can automate right-sizing recommendations directly within their CI/CD pipelines, making efficiency a default part of the development process. This allows them to scale resources dynamically and still maintain control through versioned code commits. Furthermore, sophisticated organizations use spot fleet management services to diversify instance types and availability zones, significantly reducing the risk of interruptions for even large-scale distributed workloads. To achieve comprehensive visibility, cost and performance data can be consolidated into a unified observability dashboard. This creates a single source of truth for both technical and financial stakeholders, fostering a shared culture of cloud **Financial Operations (FinOps)** where everyone is responsible for cost efficiency. A signed-off design document should precede any major architectural changes to formalize these advanced strategies.

Therefore, advanced cost optimization involves automating right-sizing in CI/CD pipelines, using fleet management for spot instances, and creating unified dashboards to foster a FinOps culture.

Related topics

This section focuses on infrastructure-level cost optimization strategies. For related topics addressed elsewhere in this book:

- Cloud migration cost considerations, see *Chapter 9, Cloud Migration and Coexistence Strategies*
- Resource governance, see *Chapter 6, Data Governance and Compliance*

Observability practices for proactive monitoring

Proactive monitoring is essential for building resilient data systems, shifting focus from reactive fixes to preventing failures before they impact users. This requires moving beyond traditional system metrics to a more expansive view of observability that includes critical metadata. For a scaled data platform, this means integrating a data catalog for asset discovery and end-to-end data lineage for impact analysis. By combining these, organizations gain deep insight into the health, quality, and interconnectedness of their data. This approach allows teams to accelerate development even as they maintain high standards of data reliability, where a version-controlled decision log file provides an immutable record of changes.

Key considerations

To establish a robust observability framework, several core principles must be addressed. These considerations form the foundation for a system that provides clarity and enables rapid, informed decision-making:

- **Unified discovery and context:** A comprehensive data catalog is the cornerstone of data democratization and governance. It provides a single source of truth for all data assets, complete with business context, ownership, and quality scores, enabling consumers to find and trust the data they need. This allows teams to innovate with new data sources while an automated tagging system ensures all assets remain discoverable and compliant.
- **End-to-end dependency mapping:** Understanding data lineage is critical for managing change and ensuring reliability in a complex ecosystem. It allows teams to perform precise impact analysis before deploying changes and accelerate root cause analysis when issues arise. Teams can confidently refactor upstream data models and still guarantee downstream stability through a pre-deployment lineage validation check.
- **Synthesized view for leadership:** Technical metrics must be translated into a consolidated view that connects operational performance to business outcomes. A unified dashboard provides leaders with key indicators like deployment frequency and rollback speed, enabling strategic oversight. Leaders can make strategic decisions quickly without sacrificing a detailed understanding of operational health, and a feature flag allows for the controlled rollout of new dashboard components.

Therefore, effective data platform observability hinges on a trusted data catalog for context, end-to-end lineage for impact analysis, and a unified dashboard for strategic oversight.

Process or steps

Creating a unified observability dashboard involves three distinct steps to correlate operational decisions with performance outcomes:

1. **Consolidate decision records:** The process begins when various deployment and management decisions create a need for a centralized audit history. To address this, all significant lifecycle events, such as product promotions or rollbacks, are captured in a structured, immutable decision log file. The result is a complete and auditable record of every critical action taken within the system.

2. **Aggregate performance metrics:** As the data product operates, it generates a continuous stream of performance data, which must be made intelligible. This involves collecting and summarizing **key performance indicators (KPIs)** for data quality, user adoption, and operational stability into a clean, aggregated performance summary file. This action produces a quantitative summary of the product's health and value over time. Distinguish between leading indicators (predictive metrics like queue depth or resource utilization that signal potential issues) and lagging indicators (outcome metrics like error rates or SLA breaches that confirm problems have occurred).

3. **Join and visualize data:** With decision logs and performance metrics available as separate sources, the final action is to synthesize them into a single, coherent view. The decision log and performance metrics files are joined on a common key, such as a deployment identifier or a relevant timestamp range, to build a comprehensive dashboard. This step results in a powerful visualization that directly links operational actions to their performance consequences.

Therefore, the process involves consolidating decision logs, aggregating performance KPIs, and then joining these sources to create a dashboard that visualizes the direct impact of actions on outcomes.

Concrete example

Consider a team managing a Fusion-Mart data product that provides sales analytics. After deploying a new version to production, the team observes an alert from their unified dashboard. The event, a **`promotion_success`**, is recorded in the decision log and appears as a marker on a timeline. Immediately following this marker, a dashboard shows a sharp decline in the **`data_quality_score`** KPI and a corresponding spike in the **`query_error_rate`** KPI, both sourced from a performance summary file. The direct visual correlation between the deployment and the negative performance impact enables the on-call engineer to bypass a lengthy investigation. They immediately trigger a rollback, an action that is also logged. The dashboard then shows the KPIs returning to their normal, healthy state. This entire sequence demonstrates the continuous improvement cycle shown in *Figure 10.1*. The team can rapidly diagnose and resolve issues while a detailed audit trail automatically documents the entire incident for post-mortem analysis, supported by a documented rollback procedure that ensures a swift response.

Therefore, a team using a unified dashboard can immediately correlate a new deployment with a drop in data quality KPIs, enabling a rapid rollback and demonstrating a closed-loop feedback system.

Cautions and limits

While a unified observability dashboard provides significant advantages, teams must be aware of potential challenges and limitations to ensure its effectiveness. These challenges are manageable with proper planning and process discipline.

First, the dashboard's insights are only as reliable as its underlying data. If the captured performance metrics are inaccurate or poorly defined, the resulting visualizations can lead to incorrect conclusions. To mitigate this, teams must implement rigorous data quality validation on the metric collection pipelines themselves. Second, it is crucial to distinguish correlation from causation. The dashboard excels at highlighting relationships between events and outcomes, but it does not prove causality. Teams should use these insights as a starting point for deeper root cause analysis, often supplemented with data lineage tools. Finally, maintaining the logging and metric aggregation systems requires an investment in automation. Teams can build a powerful monitoring system without creating excessive operational overhead by automating data ingestion pipelines. A schema definition enforced through a CI/CD pipeline prevents malformed log entries and ensures data consistency.

Therefore, teams must ensure high-quality input metrics, distinguish correlation from causation, and automate processes to manage the operational overhead of the observability system.

Glossary

Defining key terminology ensures clarity and shared understanding across technical and non-technical stakeholders:

- **Data catalog:** A centralized inventory of data assets, enriched with metadata (e.g., descriptions, ownership, quality scores) to support discovery, governance, and understanding.
- **Data lineage:** A map of the data's journey, showing its origin, transformations, and dependencies across various systems. It is essential for impact and root cause analysis.
- **KPI:** A quantifiable measure used to evaluate the success of a product or system in meeting its performance objectives.
- **Observability:** The ability to understand the internal state of a system by examining its external outputs, which for data platforms includes logs, metrics, traces, and metadata.

Defining these terms clearly helps onboard new team members quickly, even as the complexity of the data platform grows. A link to this glossary from the main observability dashboard ensures definitions are always accessible.

Expert notes

For teams seeking to advance their observability practices, several expert-level enhancements can provide deeper insights and automation. Instead of only tracking rollback frequency, advanced teams can use data lineage to quantify the blast radius of a potential failure, estimating the number of downstream reports and applications that would be affected by a change. Furthermore, the dashboard can evolve from a passive monitoring tool into an active component of a self-healing system. By defining critical KPI thresholds, teams can configure automated actions, such as triggering a rollback if data quality scores remain below a certain point for more than ten minutes post-deployment. Finally, integrating financial data provides another dimension of insight. By joining operational metrics with cloud cost data, teams can measure the cost-per-query or overall financial efficiency of a data product, identifying opportunities for performance tuning. Advanced teams can implement automated responses to system failures while a human-in-the-loop approval step prevents unintended consequences, managed through an alert routing system that directs critical notifications to a designated on-call engineer.

Therefore, experts can extend observability by quantifying blast radius, automating rollbacks based on KPI thresholds, and integrating cost metrics for financial efficiency.

Related topics

This section focuses on infrastructure-level cost optimization strategies. For related topics addressed elsewhere in this book:

- Cloud migration cost considerations, see *Chapter 9, Cloud Migration and Coexistence Strategies*
- Resource governance, see *Chapter 6, Data Governance and Compliance*

Delta Lake performance and compaction

In data lake architectures, performance is often hindered by the small file problem, where frequent data appends result in numerous small files instead of fewer large ones. This fragmentation forces query engines to manage excessive metadata and file operations, leading to significant latency. Delta Lake provides powerful tools to address this: **`OPTIMIZE`** for file compaction (the process of combining small files into larger ones) and **`Z-ORDER`** for data skipping (an efficiency technique to avoid reading unnecessary data). This discussion focuses not on command execution, but on strategically identifying when to apply these optimizations as reversible improvements, ensuring we can enhance performance while maintaining system stability through controls like a versioned commit in the delta transaction log. This data-driven approach is crucial for making informed decisions within a structured product evolution framework.

Key considerations

Understanding when and why to apply performance optimizations is critical for maintaining an efficient data lake. You can advance system performance with these tools and still preserve data integrity by using a feature flag to enable or disable the optimization job as needed. The following points are central to this strategic approach:

- The small file problem directly degrades query performance by creating excessive overhead. When a query engine must open, read, and close thousands of tiny files, it spends more time on file management than on data processing. This results in high I/O wait times and inefficient use of compute resources, slowing down analytics even when the total data volume is small.
- Delta Lake's **`OPTIMIZE`** command directly mitigates this issue through compaction. It intelligently rewrites small files into a smaller number of larger, more optimal files. This consolidation reduces metadata overhead and allows the query engine to read more data per operation, significantly improving read throughput and overall query speed.
- The **`Z-ORDER`** command further enhances performance by improving data skipping. It physically reorganizes data to co-locate related values from different columns within the same files. When queries filter on a **`Z-ORDERED`** column, the engine can use file-level statistics to skip reading entire files that do not contain relevant data, drastically reducing the data scanned and accelerating query execution.

Therefore, the small file problem creates performance-killing overhead, which **`OPTIMIZE`** solves through compaction, and **`Z-ORDER`** enhances through improved data skipping.

Process or steps

Proposing a performance optimization as a concrete, reversible improvement involves a methodical, data-driven process. This ensures that we can propose changes to improve the system and still maintain system stability with a documented rollback plan. The approach consists of three key steps:

1. **Identify performance bottlenecks:** The process begins when a performance issue is observed, such as slow-running queries or failing SLAs. This is the causal trigger, indicating that an underlying problem requires investigation. The initial step is to confirm and quantify the performance degradation using observability and monitoring tools.

2. **Analyze metadata and estimate benefits:** With a bottleneck identified, the next action is to diagnose the root cause by analyzing the Delta table's metadata. Check the number of files and their average size for the relevant partitions. If a large number of small files is confirmed, you can estimate the potential performance gain from compaction and data skipping by comparing current query times against expected improvements from reduced I/O.

3. **Propose a reversible improvement:** The final result of the analysis is a well-justified proposal to apply an optimization like **`OPTIMIZE`** or **`Z-ORDER`**. This proposal is framed as the next reversible improvement for the data product, with its benefits clearly estimated. This data-driven insight becomes a critical input for the *design* and *decision* phases of the product lifecycle, ensuring the change is both valuable and aligned with strategic goals.

Therefore, a structured process of identifying bottlenecks, analyzing metadata to estimate benefits, and proposing a reversible improvement ensures that optimizations are applied effectively.

Concrete example

Consider a data product that ingests real-time user activity events into a Delta table named **`user_events`**, partitioned by **`ingest_date`**. The streaming pipeline creates thousands of small files every hour, and analysts report that queries filtering for a specific **`user_id`** over a date range are becoming progressively slower. This allows analysts to get faster insights even as new data streams in, and system stability is assured through an A/B testing framework that validates query performance before and after the change. An investigation using **`DESCRIBE DETAIL user_events`** reveals that a single day's partition contains over 100,000 files, confirming the small file problem.

To address this, the data engineering team proposes a two-part optimization. First, they run the **`OPTIMIZE`** command to compact the small files into larger, more manageable ones. Second, because **`user_id`** is a common filter column, they apply **`Z-ORDER`** to co-locate data for the same user, dramatically improving data skipping. The SQL commands for these operations are scheduled to run daily. After implementation, query times for **`user_id`** lookups decrease from minutes to seconds, providing a measurable and significant performance improvement.

Therefore, a real-world **`user_events`** table suffering from the small file problem can be significantly accelerated by applying **`OPTIMIZE`** for compaction and **`Z-ORDER BY (user_id)`** for efficient data skipping.

Cautions and limits

While **`OPTIMIZE`** and **`Z-ORDER`** are powerful, they are not without costs and limitations. These operations are resource-intensive, consuming compute cycles and I/O to rewrite data, which can incur significant cloud infrastructure costs. Running them indiscriminately or too frequently can lead to a situation where the cost of optimization outweighs the performance benefits. It is possible to schedule these optimizations to run efficiently without impacting production workloads by setting clear resource quotas on the optimization cluster.

A critical prerequisite is to have a clearly identified performance problem before applying a solution; premature optimization can introduce unnecessary complexity and expense. Furthermore, the effectiveness of **`Z-ORDER`** depends heavily on query patterns. If filter predicates change, the existing **`Z-ORDER`** key may no longer be optimal. Therefore, optimizations should be treated as reversible improvements that are continuously evaluated against current usage patterns and performance metrics.

Therefore, **`OPTIMIZE`** and **`Z-ORDER`** are resource-intensive operations that should only be applied to solve identified bottlenecks, with their costs and benefits carefully managed.

Glossary

Key terms for understanding Delta Lake performance optimization include the following:

- **Compaction:** The process of rewriting numerous small data files into a smaller number of larger files. This reduces metadata overhead and improves read I/O efficiency.
- **Data skipping:** A query optimization technique where an engine uses file-level metadata (like min/max values for columns) to determine which files can be safely ignored during a scan, reducing the total amount of data read.
- **Small file problem:** A common performance issue in data lakes where data is stored in a large number of small files, causing significant overhead for file system operations and query planning.
- **Z-ordering:** A data layout technique that co-locates related column values within the same file or set of files. This multidimensional clustering greatly enhances the effectiveness of data skipping.

Expert notes

For experienced practitioners, the strategic application of these optimizations is key to building high-performance data products. The selection of columns for `Z-ORDER` is a critical decision; ideal candidates are high-cardinality columns frequently used in `WHERE` clauses. Applying `Z-ORDER` on a low-cardinality column like `country_code` will yield minimal benefits compared to a high-cardinality column like `user_id` or `session_id`. This allows teams to achieve higher performance from their data products and still adhere to data governance policies by using an audit trail that logs all `OPTIMIZE` and `Z-ORDER` operations.

Furthermore, these optimizations should not be viewed as one-time fixes but as part of a continuous improvement cycle for the data platform, as visualized in *Figure 10.1*. The decision to implement or adjust an optimization strategy serves as a concrete, reversible next step for a data product. This data-driven decision is a core component of the *design* and *decision* phases of a structured governance framework, ensuring that technical improvements align with the product's evolving needs and deliver measurable value.

Therefore, advanced use involves selecting high-cardinality `Z-ORDER` columns and integrating optimization tasks into a continuous improvement cycle guided by a formal decision-making framework.

Related topics

This section focuses on creating a unified observability dashboard by correlating decision logs with performance metrics. For related topics addressed elsewhere in this book:

- SLI/SLO frameworks, see *Chapter 3, Data Integration Strategy, Business Impact, and ROI*
- Audit trail implementation, see *Chapter 6, Data Governance and Compliance*

Emerging data trends

A structured methodology for designing reversible improvements is critical for the iterative development and risk mitigation of data products. This approach ensures that innovation can proceed with confidence. However, the data engineering landscape is being reshaped by emerging technologies like **large language models** (**LLM**) and the critical demands of AI governance. To maintain relevance and responsibility, data professionals must engage in continuous professional development. This ongoing learning is essential for applying established frameworks effectively, allowing teams to advance product capabilities with new AI features while maintaining rigorous compliance, all tracked within a version-controlled design document that serves as an auditable record of decisions.

Key considerations

Before integrating advanced AI, it is crucial to evaluate several key areas to ensure success. The following points outline the primary considerations for modern data product development:

- A structured framework provides the foundation for making incremental, reversible improvements to data products. This disciplined approach allows teams to experiment with new features and architectures with a clear path to revert changes if they do not yield the expected results. This methodical process ensures that progress can be made without jeopardizing the stability of the core product, with a feature flag system enabling immediate rollback of any problematic deployments.
- Emerging trends, particularly the integration of LLMs and autonomous agents, are fundamentally altering data engineering practices. These technologies impact the entire data lifecycle, from ingestion and processing to the features delivered to end-users. Understanding these shifts is vital for designing data pipelines and products that are not only powerful but also efficient and scalable for future demands.
- The evolving field of AI governance introduces essential new considerations for ethical design, regulatory compliance, and responsible deployment. Integrating these principles from the outset is no longer optional; it is a core requirement for building trust with users and stakeholders. We can build these sophisticated systems and still ensure they operate within ethical boundaries by embedding governance checks throughout the development process.

Therefore, a successful data product strategy balances a structured, reversible development methodology with a forward-looking approach to new technologies like LLMs and a foundational commitment to AI governance.

Process or steps

Adapting a data product to incorporate AI advancements while ensuring it remains robust and compliant involves three key steps:

1. **Analyze impact and scope:** The initial step is to conduct a thorough analysis of how a new technology, such as an LLM, will affect the existing data architecture and product features. This involves identifying potential data sources, defining necessary transformations, and scoping the reversible improvement to ensure it is manageable. This analysis allows teams to advance their technical capabilities without introducing unforeseen risks to the system.

2. **Integrate governance frameworks:** With a clear understanding of the technical impact, the next step is to embed relevant AI governance principles directly into the project plan. This includes defining ethical guidelines, establishing compliance checks, and planning for human oversight. An automated alert from a unified observability dashboard can serve as a key assurance artifact, notifying the team of any deviations from established norms.

3. **Implement and monitor iteratively:** The final step is to implement the change within the structured development framework, deploying it incrementally. Using a continuous observability loop is crucial for monitoring the feature's performance, user impact, and adherence to governance standards. This data-driven feedback allows for rapid, informed iterations, ensuring the final product is both effective and responsible.

Therefore, the process for evolving data products requires a sequence of analyzing technological impact, integrating governance, and using observability loops for iterative implementation and monitoring.

Concrete example

Let us consider a practical example of applying this process to the Fusion-Mart data product by introducing an LLM-powered feature for generating product summaries. The team begins by assessing how the LLM will interact with their **`product_catalog`** data and what new data quality checks are needed for the generated text. They then integrate governance by defining rules to prevent biased language and establishing a workflow for human review of the AI-generated summaries. The feature is subsequently deployed behind a feature flag to a small subset of users, allowing the team to test this innovative capability in a live environment even as they retain full control. An observability dashboard tracks summary quality and user feedback, enabling rapid adjustments. The specific commit hash in the code repository provides a concrete assurance artifact, enabling a precise rollback to the previous state if the feature underperforms.

Therefore, implementing an LLM-based product summary generator for Fusion-Mart demonstrates how to assess impact, embed governance, and use a feature-flagged, observable deployment to innovate safely.

Cautions and limits

Integrating advanced AI technologies introduces unique challenges that require careful management. A primary concern is model drift (a phenomenon where model performance degrades over time), which necessitates continuous monitoring and periodic retraining to ensure accuracy. Furthermore, the landscape of AI regulation is new and constantly evolving, requiring systems to be flexible enough to adapt to future compliance demands. Teams can navigate this uncertainty while building robust systems by maintaining a close watch on regulatory developments. Perhaps the most significant prerequisite is addressing potential skill gaps within the team, as expertise in LLM operations and AI ethics is crucial. A commitment to continuous professional development, which can lead to career progression from a technical to a strategic role, as shown in *Figure 10.2*, is the most effective mitigation. A mandatory pre-deployment checklist requiring sign-off from a data governance expert acts as a practical safeguard.

Therefore, the key challenges in AI integration include managing model drift, navigating regulatory uncertainty, and addressing skill gaps through continuous professional development.

The following figure shows the career progression from technician to a data governance strategist:

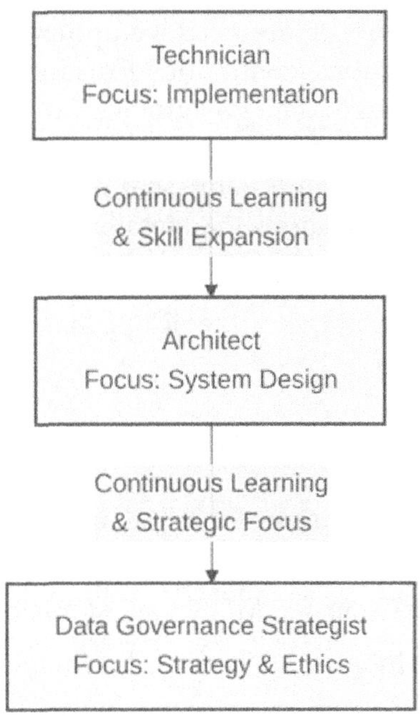

***Figure 10.2**: Career progression from technician to a data governance strategist*

Glossary

The following are the key terms that have been discussed in this section:

- **AI governance:** A framework of rules, practices, and processes for ensuring that artificial intelligence systems are developed and used responsibly, ethically, and in compliance with regulations.
- **LLM:** A type of artificial intelligence model trained on vast amounts of text data to understand and generate human-like language.
- **Observability loop:** A continuous process of monitoring a system's outputs, collecting data, and using that data to inform improvements and decisions.
- **Reversible improvement:** A change or enhancement to a system that is designed to be easily undone or rolled back to a previous state without causing significant disruption.

Expert notes

For experienced practitioners, moving beyond basic implementation to achieve excellence requires a holistic perspective. This involves establishing a unified observability strategy, to gain a comprehensive view of system performance and compliance. This approach feeds into a broader continuous improvement cycle, similar to the one depicted in *Figure 10.1*, where insights from operations directly inform the next development iteration. This strategic mindset also has significant career implications; continuous learning enables professionals to evolve from technical implementation roles to data governance strategists, as illustrated in the career progression model in *Figure 10.2*. By adopting this advanced viewpoint, experts can drive significant innovation and still maintain operational stability, supported by a documented incident response plan for AI model failures that details precise rollback procedures.

Therefore, advanced practitioners should focus on unified observability and continuous improvement cycles, recognizing that this strategic approach also drives professional growth into governance and strategy roles.

Related topics

This section focuses on the strategic framework for integrating AI into data products. For related topics addressed elsewhere in this book:

- AI governance gates, see *Chapter 7, MLOps for AI Model Deployment and Monitoring*
- Four-Gate Governance Model, see *Chapter 1, Evolution of Data Architecture*

Lab: Scaling and optimizing tiny warehouses

In this lab, you will apply the scaling and optimization techniques covered in this chapter using a small-scale data warehouse environment. Working with tiny warehouses, you will implement performance tuning, cost optimization, and observability practices in a controlled setting that demonstrates how these techniques scale from proof-of-concept to production environments.

Lab environment: All labs run in GitHub Codespaces with a dev container. This lab demonstrates the practical application of optimization strategies using a tiny warehouse that mirrors real-world constraints.

This lab produces `labs/ch10/artifacts/result.json` containing optimization results, performance metrics before and after tuning, cost analysis, and recommendations for scaling these techniques to larger environments. This structured output demonstrates how small-scale experiments inform production optimization strategies.

Note: **The lab uses accept/reject for simplicity. In the book text, we use Go/Retry/Kill to describe the same decision workflow.**

Conclusion

This chapter equipped you with essential skills for scaling data platforms through optimization and proactive monitoring. You learned to identify and resolve performance bottlenecks through systematic analysis, optimize cloud costs by balancing performance with resource efficiency, and implement observability practices that enable proactive issue detection through SLIs and automated alerting.

Through the hands-on lab, you applied scaling and optimization techniques using a tiny warehouse environment, implementing performance tuning, cost optimization, and observability practices in a controlled setting. This practical experience demonstrates how to systematically identify bottlenecks, apply targeted improvements, and validate results, essential skills for managing data platforms at any scale.

Beyond technical patterns, you learned how DevOps practices enable successful platform scaling. Performance tuning, cost optimization, and observability are not one-time activities but continuous practices integrated into development workflows through automation, monitoring, and iterative improvement. These practices ensure your data platform evolves efficiently while maintaining reliability and controlling costs.

The optimization and governance techniques covered in this chapter, and throughout this book, provide a foundation for managing increasingly complex data ecosystems. As your platform grows, these practices become even more critical, whether expanding to new domains, integrating additional data sources, or scaling to support larger user bases. By combining the DevOps automation, governance standards, and optimization techniques you have learned, you're now equipped to build and maintain data platforms that evolve efficiently while remaining reliable and cost-effective.

Fusion-Mart's optimized data platform now delivers sub-minute query responses while reducing cloud costs by 40%, proving that performance and efficiency can scale together.

Index

D

E

F

G

H

I

Made in the USA
Las Vegas, NV
21 March 2026

43806488R00138